IEE POWER AND ENERGY SERIES 31

Series Editors: Professor A. T. Johns
D. F. Warne

EMBEDDED GENERATION

Other volumes in this series:

EMBEDDED GENERATION

Nick Jenkins, Ron Allan,
Peter Crossley, David Kirschen
and Goran Strbac

The Institution of Electrical Engineers

Published by: The Institution of Electrical Engineers, London,
United Kingdom

© 2000: The Institution of Electrical Engineers

The Institution of Electrical Engineers,
Michael Faraday House,
Six Hills Way, Stevenage,
Herts. SG1 2AY, United Kingdom

British Library Cataloguing in Publication Data

A CIP catalogue record for this book
is available from the British Library

ISBN 0 85296 774 8

Printed in England by TJ International Ltd., Padstow, Cornwall

Contents

Preface

Power systems were originally developed in the form of local generation supplying local demands, the individual systems being built and operated by independent companies. During the early years of development, this proved quite sufficient. However, it was soon recognised that an integrated system, planned and operated by a specific organisation, was needed to create an effective system that was both reasonably secure and economic. This led to centrally located generation feeding the demands via transmission and distribution systems. In the event, a significant amount of local generation was left isolated within the developing distribution systems, but this was gradually mothballed and subsequently decommissioned so that, by the 1970s, most of it had disappeared from the UK electricity supply industry. This trend may well have continued but for the need to minimise energy use, particularly that believed to create environmental pollution. Consequently, governments and energy planners have more recently been actively developing alternative and cleaner forms of energy production, these being dominated by renewables (wind, solar, etc.), local CHP plant and the use of waste products. Paradoxically, economics and the location of the fuel and/or energy sources have meant that these newer sources have had to be mainly connected into distribution networks rather than at the transmission level. A full circle has therefore evolved with generation being 'embedded' in distribution systems and 'dispersed' around the system rather than being located and dispatched centrally or globally.

Over several decades, models, techniques and application tools have developed that recognised the central nature of generation, and there are many excellent texts that relate to and describe the assessment of such systems. However, some very specific features of embedded generation, namely that relatively small amounts of generation are dispersed around the system, often connected into relatively weak networks, and that the generators are not usually dispatched by the network operator has meant that existing techniques and practices have had to be reviewed and updated to take these features into account. This book addresses these features and issues.

The Electrical Energy and Power Systems Group in The Manchester Centre of Electrical Energy at UMIST have considerable expertise and experience in all aspects of the problems associated with embedded generation. We were therefore very pleased to have been invited by the IEE to write this text. We have taken great care not to create a book that would be perceived to be specifically research related, which in a rapidly developing activity would be all too easy. Instead we have attempted to create a text that addresses all the main issues, that places these issues into the context of real system applications and that describes current practices and applications. However, we recognise that these practices are continually changing, so we have also addressed various ways forward and the possible solutions to problems that have not yet been resolved. Obviously, we are not in a position to say whether all these objectives have been accomplished; only the reader can do this. However, we sincerely trust that they have been.

Finally, no book, including this, could be produced in isolation, and we are indebted to all those colleagues and individuals with whom we have been involved in industry and in an extensive range of professional organisations. Particular thanks are due to former UMIST colleagues for the use of their IPSA program for the examples in Chapters 3 and 4. The other authors would also like to express their affection and thanks to Ron Allan for his leadership of the Group and for building the team which resulted in this book.

Contributors

Nick Jenkins joined UMIST in 1992 and was appointed Professor of Electrical Energy and Power Systems in 1998. His previous career included 14 years industrial experience, of which 5 years were in developing countries. He has worked for Wind Energy Group, BP Solar and Ewbank and Partners on both conventional and renewable power systems. His present research interests include renewable energy and embedded generation. During 1998/99 he was the secretary of a CIRED international working group on dispersed generation.

Ron Allan is Professor of Electrical Energy Systems at UMIST. His interests centre on assessing power system reliability and customer worth of supply. He has authored/co-authored over 200 technical papers and three reliability textbooks. As a result of his research, he was awarded a DSc in 1987 and the IEE Institution Premium in 1996. He is a Fellow of the IEE, the IEEE and the Safety and Reliability Society, and is currently Deputy Chairman of the IEE Power Division.

Peter Crossley joined UMIST in 1991 where he is currently a Reader specialising in Power System Protection and Control. His previous career included 13 years with ALSTOM T&D where he was involved in the design and application of protection relays. He graduated from UMIST in 1977 with a BSc degree and the University of Cambridge in 1983 with a PhD degree. He has authored/co-authored over 100 technical papers and is a member of various IEE, IEEE and CIGRE committees on Protection. In 2001 he will be the Chairman of the IEE Developments in Power System Protection Conference.

Daniel Kirschen received his Mechanical and Electrical Engineer's Degree from the Free University of Brussels, Belgium in 1979. With financial assistance from the Belgian American Educational Foundation, he began postgraduate studies at the University of Wisconsin-Madison, where he received his MSc and PhD degrees in Electrical Engineering in 1980 and 1985 respectively. He then joined Control Data Corporation's Energy Management Systems Division where he worked on the development of network optimisation and artificial intelligence software for electric utilities. In 1994 he took an academic position at UMIST where he is currently a Reader.

Goran Strbac joined UMIST in November 1994 where he is a Reader. His previous career included 4 years industrial experience and more than 6 years research and teaching experience. His research interests are in power systems operation and development in a competitive environment. He has been involved in a number of collaborative projects on the economics of embedded generation and pricing of network and ancillary services. He is the convenor of a CIGRE working party on 'Economic interaction between embedded generation and the power system'.

Glossary

AENS: Average energy not supplied or served, i.e. energy not supplied to the average customer

Anti-islanding protection: Electrical protection to detect if a section of the distribution network has become isolated or 'islanded'. The embedded generator is tripped to prevent continuous operation of the autonomous power 'island'. Effectively synonymous with loss-of-mains protection

ASAI: Average system availability index, i.e. the ratio of the total number of customer hours that service was available during a year to the total number of customer hours demanded

ASUI: Average system unavailability index, i.e. the ratio of the total number of customer hours that service was unavailable during a year to the total number of customer hours demanded

Asynchronous generator: Synonymous with induction generator. The rotor runs at a slightly faster rotational speed than the stator field. The construction of an asynchronous (or induction) generator is very similar to that of an induction motor of similar rating

Auto-reclose: The automatic reclosure of a circuit breaker after it has opened when a fault has been detected

Biomass: Biological material used as fuel for conversion to heat or electrical energy

CAIDI: Customer average interruption duration index, i.e. average interruption duration per customer interruption

CAIFI: Customer average interruption frequency index, i.e. average number of interruptions per customer affected per year

CHP: Combined heat and power

CIRED: International Conference on Electricity Distribution Networks

Combined Heat and Power: The simultaneous production of heat and electrical energy. Synonymous with co-generation. Industrial combined heat and power systems typically produce hot water or steam and electrical power for use within the host site. The heat output from combined heat and power plants may also be used for district heating

Despatched: Generating plant which is under central control and so 'despatched' or controlled by the power system operator

Dispersed generation: Synonymous with embedded generation

EENS: Expected energy not supplied

EG: Embedded generation

EIR: Energy index of reliability, i.e. LOEE normalised by dividing by the total energy demanded

EIU: Energy index of unreliability, i.e. (1-EIR)

Embedded generation: Generation which is connected to or 'embedded in' the distribution network. Synonymous with dispersed generation

ENS: Energy not supplied

Fault level: The fault level at a given point in a power network is a measure of the magnitude of the fault current that would result from a balanced three-phase fault at that point. The fault level is higher for networks that are more heavily meshed and increases as the point considered moves closer to generators

Flicker: Used to describe high frequency (up to 10 Hz) variations in network voltage magnitude which may give rise to noticeable changes in light intensity or 'flicker' of incandescent lamps

Fuel cells: An electrolyte cell supplied continuously with chemical material, stored outside the cell, which provides the chemical energy for conversion to electrical energy

Generator reserve requirements: Generation reserve is used to balance generation and demand following the unexpected breakdown of plant or after a demand forecast error

Geothermal plant: Generation plant which uses heat from the earth as its input

Grid supply point: A grid supply point is a substation where the transmission network is connected to the sub-transmission or distribution network. In the United Kingdom, this connection is usually made by transformers that step the voltage down from 400kV or 275kV to 132 kV

Guaranteed Standards: These are service Standards set by the UK Regulator (OFGEM, formerly OFFER) which must be met by all Public Electricity Suppliers (PESs). If the PES fails to provide the level of service required, it must make a payment to the customer affected. The level of each Standard and the level of payment to be made are reviewed regularly by the Regulator, and have been continuously updated since their introduction in July 1991

Harmonic distortion: Distortion of the network voltage or current from a true sinusoid

Induction generator: Synonymous with asynchronous generator

Intertie: A line or group of lines that connect two power systems that are operated by different companies

LOEE: Loss of energy expectation, i.e. the expected energy that will not be supplied due to those occasions when the load exceeds the available generation. It encompasses the severity of the deficiencies as well as their likelihood. Essentially the same as EENS or similar terms such as expected unsupplied (unserved) energy (EUE)

LOLE: Loss of load expectation, i.e. the average number of days on which the daily peak load is expected to exceed the available generating capacity. Alternatively it may be the average number of hours during which the load is expected to exceed the available capacity. It defines the likelihood of a deficiency but not the severity, nor the duration

LOLP: Loss of load probability, i.e. the probability that the load will exceed the available generation capacity. It defines the likelihood of a deficiency but not the severity

Loss-of-mains protection: Electrical protection applied to an embedded generator to detect loss of connection to the main power system. Synonymous with anti-islanding protection

Neutral grounding: The connection of the neutral point of a three-phase power system to ground (or earth)

Neutral voltage displacement: Electrical protection relay technique used to measure the displacement of the neutral point of a section of the power system. Used particularly to detect earth faults on networks supplied from a delta-connected transformer winding

Permanent outage: An outage associated with damaged faults which require the failed component to be repaired or replaced

Photovoltaic: The physical effect by which light is converted directly into electrical energy

Power: Electrical power is the product of a current and a voltage. In ac circuits, this product is called *apparent power*. The angle difference between the current and voltage waveforms is called the *phase angle* and the cosine of this phase angle is called the *power factor*. The *real power* is the electrical power that can be transformed into another form of energy. It is equal to the product of the apparent power and the power factor. *Reactive power* is not transformed into useful work and is a mathematical construct used to represent the oscillation of power between inductive and capacitive elements in the network. It is equal to the product of the apparent power and the sine of the phase angle. A purely active load (i.e. a load such as an electric heater that consumes only real power) has a power factor of 1.0. Practical loads usually have a power factor smaller than 1.0. A load with a power factor of 0.0 would be purely reactive and would not transform any electrical energy into practical work

Quality of supply: Perfect 'quality of supply' in the most general sense means an undistorted waveform without any interruptions of any duration. Some organisations however associate quality only with waveform distortions, e.g. harmonics, voltage sags, etc. Others associate it with short and/or long interruptions of supply

Radial distribution feeder: A distribution feeder (underground cable or overhead line) which is connected to one supply point only

Rate of change of frequency (ROCOF) – an electrical protection technique based on the rate of change of electrical frequency used in loss-of-mains or anti-islanding relays

Reliability: This is an inherent attribute of a system, which is characterised by a specific group of measures that describe how well the system performs its basic function of providing customers with a supply of energy on demand and without interruption

ROCOF: See rate of change of frequency

SAIDI: System average interruption duration index, i.e. average interruption duration per customer served per year

SAIFI: System average interruption frequency index, i.e. average number of interruptions per customer served per year

Scheduled maintenance outage: An outage that is planned in advance in order to perform preventive maintenance

Stability: Under normal circumstances, the power network is capable of handling the flow of power from generators to loads. Following unpredictable events, such as faults or failures, the network may lose its ability to transfer this power. Under such circumstances, the power system is said to have lost its stability. Stability can be lost either through transient instability or voltage instability. Transient instability means that, following a fault, the rotor angle of one or more generators increases uncontrollably with respect to the angle of the other generators. Voltage instability means that, due to a lack of reactive power generation near the loads, the voltage in parts of the system collapses uncontrollably

Stirling engine: An external combustion engine based on the Stirling thermodynamic cycle. Stirling engines are of considerable potential significance for small-scale CHP schemes

Symmetrical components: Symmetrical components are a way of describing the voltages and currents in a three-phase system that is particularly useful when studying unbalanced conditions such as some kinds of faults

Synchronous generator: A generator whose rotor operates in synchronism with its stator field. In its usual construction a synchronous generator allows independent control of real and reactive output power

System minutes: The value of LOEE normalised by dividing by the peak demand. It is a normalised value of energy and does not represent the duration of the deficiency unless all the energy was interrupted at the time of peak demand, which is extremely unlikely

Temporary outage: An outage associated with undamaged faults, which are restored by manual switching, fuse replacement or similarly lengthy restoration times

Transient outage: An outage associated with undamaged faults, which are restored by automatic switching

Transient voltage variations: Rapid (greater than 0.5 Hz) changes in the magnitude of the network voltage. May be repetitive or may refer to a single event

Voltage vector shift: A protective relay technique based on measuring the change or shift of the angle of the system voltage vector. The technique is used in loss-of-mains relays

Chapter 1

Introduction

1.1 Embedded or dispersed generation

Modern electrical power systems have been developed, over the last 50 years, mainly following the arrangement indicated in Figure 1.1. Large central generators feed electrical power up through generator transformers to a high voltage interconnected transmission network. The transmission system is used to transport the power, sometimes over considerable distances, which is then extracted from the transmission network and passed down through a series of distribution transformers to final circuits for delivery to the customers. However, recently there has been a considerable revival in interest in connecting generation to the distribution network and this has come to be known as embedded or dispersed generation. The term 'embedded generation' comes from the concept of generation embedded in the distribution network while 'dispersed generation' is used to distinguish it from central generation. The two terms can be considered to be synonymous and interchangeable.

Between 1997 and 1999 dispersed generation was investigated by working groups of both CIGRE (The International Conference on Large High Voltage Electric Systems) and CIRED (The International Conference on Electricity Distribution Networks). The two reports of these Working Groups [1,2], which are largely complementary, provide the most comprehensive review available of both the extent and main issues associated with dispersed generation and, together with supporting documents, are reviewed in the first part of this chapter.

There is, at present, no universally agreed definition of what constitutes embedded or dispersed generation and how it differs from conventional or central generation. Some common attributes of embedded or dispersed generation may be listed as [1,3]:

- not centrally planned (by the utility)
- not centrally despatched

- normally smaller than 50–100 MW
- usually connected to the distribution system.

The distribution system is taken to be those networks to which customers are connected directly and which are typically of voltages from 230/400 V up to 145 kV.

In some countries a strict definition of embedded generation is made, based either on the rating of the plant or on the voltage level to which the embedded generation is connected. However these definitions usually follow from the extent of the authority of particular, national, technical documents used to specify aspects of the connection or operation of embedded generation and not from any basic consideration of its impact on the power system. This book is concerned with the fundamental behaviour of generation connected to distribution systems and so a rather broad description of embedded or dispersed generation is preferred to any particular limits based on plant size, voltage or prime mover type.

1.2 Reasons for embedded generation [1,2,4]

The conventional arrangement of a modern large power system (illustrated in Figure 1.1) offers a number of advantages. Large generating units can be made efficient and operated with only a relatively small number of personnel. The interconnected high voltage transmission network allows generator reserve requirements to be minimised and the most efficient generating plant to be despatched at any time, and

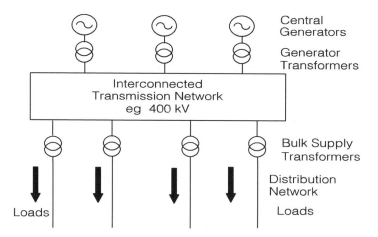

Figure 1.1   Conventional large electric power system

bulk power can be transported large distances with limited electrical losses. The distribution networks can be designed for unidirectional flows of power and sized to accommodate customer loads only. However, over the last few years a number of influences have combined to lead to the increased interest in embedded generation schemes. The CIRED survey [2] asked representatives from 17 countries what were the policy drivers encouraging embedded generation. The answers included:

- reduction in gaseous emissions (mainly CO_2)
- energy efficiency or rational use of energy
- deregulation or competition policy
- diversification of energy sources
- national power requirement.

The CIGRE report [2,4] listed similar reasons but with additional emphasis on commercial considerations such as:

- availability of modular generating plant
- ease of finding sites for smaller generators
- short construction times and lower capital costs of smaller plant
- generation may be sited closer to load, which may reduce transmission costs.

Environmental impact is a major factor in the consideration of any electrical power scheme, and there is a generally accepted concern over gaseous emissions from fossil-fuelled plant. As part of the Kyoto Protocol, both the EU and the UK have to reduce substantially emissions of CO_2 to help counter climate change. Hence most governments have programmes to support the exploitation of the so-called new renewable energy resources, which include wind power, micro-hydro, solar photovoltaics, landfill gas, energy from municipal waste and biomass generation. Renewable energy sources have a much lower energy density than fossil fuels and so the generation plants are smaller and geographically widely spread. For example, wind farms must be located in windy areas, while biomass plants are usually of modest capacity due to the cost of transporting fuel with relatively low energy density. These small plants, typically of less than 50 MW in capacity, are then connected into the distribution system. In many countries the new renewable generation plants are not planned by the utility but are developed by entrepreneurs and are not centrally despatched but generate whenever the energy source is available.

Cogeneration or Combined Heat and Power (CHP) schemes make use of the waste heat of thermal generating plant for either industrial processes or space heating and are a well established way of increasing overall energy efficiency. Transporting the low temperature waste heat from thermal generation plants over long distances is not economical and so it is necessary to locate the CHP plant close to the heat load. This again leads to relatively small generation units, geographically distributed and with their electrical connection made to the distribution network. Although CHP units can, in principal, be centrally despatched they tend to be operated in response to the requirements of the heat load or the electrical load of the host installation rather than the needs of the public electricity supply.

The commercial structure of the electricity supply industry is also playing an important role in the development of embedded generation. In general a deregulated environment and open access to the distribution network is likely to provide greater opportunities for embedded generation, although Denmark provides an interesting counterexample where both wind power and CHP have been extensively developed in a vertically integrated ownership structure.

Finally, in some countries the fuel diversity offered by embedded generation is considered to be valuable while in some developing countries the shortage of power is so acute that any generation is to be welcomed.

At present, embedded generation is seen almost exclusively as producing energy (kWh) and making no contribution to other functions of the power system (e.g. voltage control, network reliability, generation reserve capacity, etc.). Although this is partly due to the technical characteristics of the embedded generation plant this restricted role of embedded generation is predominantly caused by the administrative and commercial arrangements under which it presently operates. There are already examples of embedded generation used to provide generation reserve (Électricité de France can call upon some 610 MW of despatchable, distributed diesel generators), while in the UK there are isolated examples of diesel generators being used to provide reinforcement of the distribution network where it is not possible to obtain planning permission for the construction of additional overhead circuits.

Looking further into the future, the increased use of fuel cells, micro-CHP using novel turbines or Stirling engines and photovoltaic devices integrated into the fabric of buildings may all be anticipated as possible sources of power for embedded generation. If these technologies become cost-effective then their widespread implementation will have very considerable consequences for existing power systems.

1.3   Extent of embedded generation

Table 1.1 shows the extent of embedded generation in countries which responded to the CIRED survey [2], while Figure 1.2 compares the data with that of the CIGRE [1] and EnergieNed [5] reports. (EnergieNed is an association of energy distribution companies in the Netherlands.) Some care is necessary in interpreting the rather inconsistent data as

Table 1.1 Extent of embedded generation in MW [2]

MW	Despatched, diesel and GT	Cogen., not despatched	Wind	Steam	Hydro	PV	Other (inc. waste)	Total of dispersed generation	System installed capacity	System maximum demand	% Dispersed generation/installed capacity
Australia	718	1747	5	2754				5224	42 437	29 841	12.3
Austria*		70	13.3		616	0.7		700			4
Belgium	214	1174	5		97			1938	14 693	11 972	13.2
Czech Republic	977						1938	1913			
Denmark		2000	1450					3450	12 150	6400	28
France**	610	435	8		450		250	1753	114 500	68 900	2.5
Germany†		2800	1545		3333	17	904	8599			
Greece		3	40					43	9859	6705	0.4
India			970		155	32		1300			
Italy	492 (not desp.)	766	34		2159	5	252	3708	70 641	43 774	5.2
Netherlands		4736	427		37		80	5280	18 981	12 000	28
Poland		3000			2008			5008	33 400	23 500	15
Spain		2500			1500 (all renew)			4000	50 311	27 251	8
UK		3732	330		1494		421	5977	68 340	56 965	8.7

* Plants up to 10 MW rating only
** Plants connected at voltage levels up to 20 kV only
† Includes utility owned plant

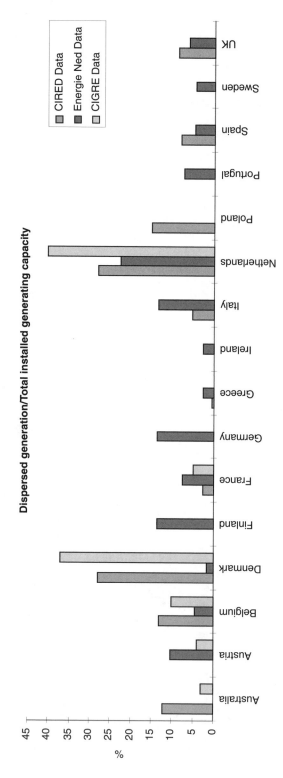

Figure 1.2   Extent of embedded generation [1,2,5]

some countries use a strict definition of embedded generation while others are more inclusive, and there are obvious discrepancies between the data for the same countries. However, a number of general conclusions may be drawn.

It can be seen that the extent of embedded generation in modern power systems is significant, with some 48 GW of installed capacity in the 14 countries shown. Although this may be only a modest fraction of the peak demand, in terms of system operation the level of penetration of embedded generation at times of minimum system load is more significant. There are already recorded instances of distribution utilities becoming net exporters of power during periods of low load because of the presence of large amounts of embedded generation. Figure 1.3 shows a comparison of dispersed generation and consumption in Denmark where a combination of distributed CHP and wind farms can already exceed the system load during periods of low customer demand. The projections for dispersed generation capacity in 2005 and 2015 are obviously predictions but are based on Danish government policy.

Table 1.1 shows that CHP (cogeneration) is an important form of embedded generation with some 40% of the installed capacity. This reflects the large gains in energy efficiency possible with CHP and hence the attractive economic performance of these schemes. Of the new renewable forms of generation, small hydro is most significant, with some 20% of the installed capacity, followed by wind with nearly 10%. Biomass and waste to energy make only a modest contribution in these countries, while solar photovoltaics has yet to be widely introduced.

It is interesting to note that, in some countries, some of the embedded

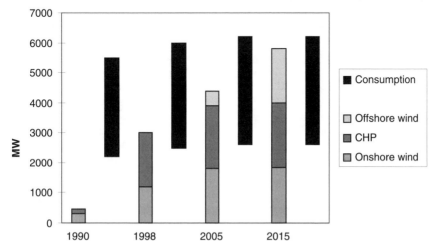

Figure 1.3   Dispersed generation and consumption in Denmark (after [1] and personal communication)

generation plant is, in fact, under central control and despatched. Even where the embedded generation is not despatched, large embedded generating stations (mainly CHP plants) may be obliged to inform the utility of their generating schedule a week in advance.

A final general conclusion is that the type and extent of embedded generation differs very significantly from country to country. This may partly be explained by some countries having particular renewable energy resources or, for example, others having large chemical complexes with CHP schemes. However, the differences also arise due to the very different policy and administrative approaches adopted which then strongly influence if a particular technology becomes commercially attractive.

The rate of development of the different forms of embedded generation varies. At present the fastest growing of the new renewable technologies is wind energy, with some 12 GW of capacity installed throughout the world (Table 1.2). The European Union has an aggressive agenda to encourage the introduction of new renewable energy schemes and Table 1.3 is taken from a White Paper issued by the European Commission in November 1997 [6]. Although not official policy of the European Union, the White Paper gives useful insights into where a significant increase in the contribution to bulk energy production might be made. It may be seen that the main contenders are wind energy and biomass. Future hydro-capacity is limited by the availability of sites, and photovoltaic generation is likely to remain expensive and with a relatively small total capacity during the time-scales envisaged. The European geothermal resource is small but biomass is clearly an important technology, although its development will be linked to European agricultural policy. A significant part of the expected growth in wind farms is likely to occur offshore. Denmark has already declared a policy of installing 4000 MW of offshore wind turbine capacity by 2030 and is

Table 1.2   Installed wind turbine capacity (MW) in January 2000

Germany	3 899
Denmark	1 761
Spain	1 131
UK	351
Total Europe	*8 349*
North America	2 617
India	1 062
Other	427
World total	12 455

Source: *Windpower Monthly* news magazine

Table 1.3 *Estimated contribution of renewables to European Union electrical energy supply [6, 7]*

Energy source	1995 TWh	2010 TWh
Wind	4	80
Hydro	307	355
– large (⩾ 10 MW)	270	300
– small (<10 MW)	37	55
Solar photovoltaic	0.03	3
Biomass	22.5	230
Geothermal	3.5	7

commencing with a programme of 750 MW. The UK government has yet to announce its intentions but is likely to require 1000–2000 MW of installed capacity by 2010 if its declared target of providing 10% of UK electricity from renewables by that date is to be achieved. The UK position on renewables has recently been comprehensively reviewed in a report from the House of Lords [7].

1.4 Issues of embedded generation

Modern distribution systems were designed to accept bulk power at the Bulk Supply Transformers and to distribute it to customers. Thus the flow of both real power (P) and reactive power (Q) was always from the higher to the lower voltage levels. This is shown schematically in Figure 1.4 and, even with interconnected distribution systems, the behaviour of the network is well understood and the procedures for both design and operation long established.

However, with significant penetration of embedded generation the power flows may become reversed and the distribution network is no longer a passive circuit supplying loads but an active system with power flows and voltages determined by the generation as well as the loads. This is shown schematically in Figure 1.5. For example, the Combined Heat and Power (CHP) scheme with the synchronous generator (S) will export real power when the electrical load of the premises falls below the output of the generator but may absorb or export reactive power depending on the setting of the excitation system of the generator. The wind turbine will export real power but is likely to absorb reactive power as its induction (sometimes known as asynchronous) generator (A) requires a source of reactive power to operate. The voltage source convertor of the photovoltaic (pv) system will allow export of real power at

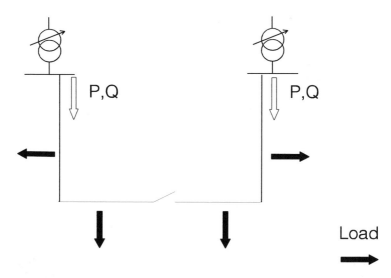

Figure 1.4   Conventional distribution system

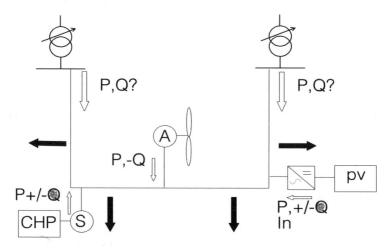

Figure 1.5   Distribution system with embedded generation

a set power factor but may introduce harmonic currents, as indicated in
Figure 1.5. Thus the power flows through the circuits may be in either
direction depending on the relative magnitudes of the real and reactive
network loads compared to the generator outputs and any losses in the
network.

The change in real and reactive power flows caused by embedded generation has important technical and economic implications for the power
system. To date, most attention has been paid to the immediate technical

issues of connecting and operating generation on a distribution system, and most countries have developed standards and practices to deal with these. In general, the approach adopted has been to ensure that any embedded generation does not reduce the quality of supply offered to other customers and to consider the generators as 'negative load'. The economic consequences and opportunities of embedded generation are only now being considered, and it is likely that these will become apparent most quickly in electricity supply industries which are deregulated and where there is a clear distinction between electricity supply (i.e. provision of kWh) and electricity distribution (i.e. provision of distribution network services).

1.5 Technical impacts of embedded generation on the distribution system

In this section the main technical impacts of embedded generation on the distribution system are reviewed. Later chapters of the book will deal with these topics in more detail, but the intention here is to provide an introductory overview of the issues.

1.5.1 Network voltage changes

Every distribution utility has an obligation to supply its customers at a voltage within specified limits. This requirement often determines the design and expense of the distribution circuits and so, over the years, techniques have been developed to make the maximum use of distribution circuits to supply customers within the required voltages. The voltage profile of a radial distribution feeder is shown in Figure 1.6.

The precise voltage levels used differ from country to country but the principle of operation of radial feeders remains the same. Table 1.4 shows the normal voltage levels used.

Figure 1.6 shows that the ratio of the MV/LV transformer has been adjusted using off-circuit taps so that at times of maximum load the most remote customer will receive acceptable voltage. During minimum load the voltage received by all customers is just below the maximum allowed. If an embedded generator is now connected to the end of the circuit then the flows in the circuit will change and hence the voltage profile. The most onerous case is likely to be when the customer load on the network is at a minimum and the output of the embedded generator must flow back to the source. It will be shown in Chapter 3 that for a lightly loaded distribution network the approximate voltage rise (ΔV) due to the generator is given (in per unit) by

$$\Delta V = (PR + XQ)/V \tag{1.1}$$

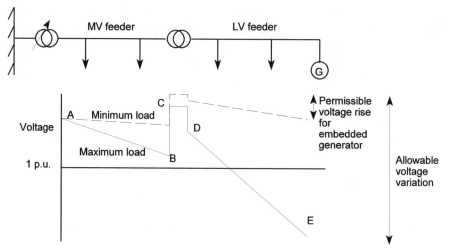

Figure 1.6   Voltage variation down a radial feeder (after Reference 8)

A   voltage held constant by tap-changer of distribution transformer
A–B   voltage drop due to load on MV feeder
B–C   voltage boost due to taps of MV/LV transformer
C–D   voltage drop in MV/LV transformer
D–E   voltage drop in LV feeder

Table 1.4   Voltage levels used in distribution circuits

	Definition	Typical UK voltages
Low voltage (LV)	LV <1 kV	230/400 V
Medium voltage (MV)	1 kV <MV <50 kV	33 kV, 11 kV
High voltage (HV)	50 kV <HV <220 kV	132 kV

where P = active power output of the generator
        Q = reactive power output of the generator
        R = resistance of the circuit
        X = inductive reactance of the circuit
        V = nominal voltage of the circuit

In some cases, the voltage rise can be limited by reversing the flow of reactive power (Q) either by using an induction generator or by underexciting a synchronous machine and operating at leading power factor. This can be effective on MV overhead circuits, which tend to have a higher ratio of X/R. However, on LV cable distribution circuits the dominant effect is that of the real power (P) and the network resistance (R) and so only very small embedded generators may generally be connected out on LV networks.

For larger generators a point of connection is required either at the LV busbars of the MV/LV transformer or, for larger plants, directly to an MV or HV circuit. In some countries simple design rules are used to give an indication of the maximum capacity of embedded generation which may be connected at different points of the distribution system. These simple rules tend to be rather restrictive and more detailed calculations often show that more generation can be connected with no difficulties. Table 1.5 shows some of the rules that are used.

An alternative simple approach to deciding if a generator may be connected is to require that the three-phase short-circuit level (fault level) at the point of connection is a minimum multiple of the embedded generator rating. Multiples as high as 20 or 25 have been required for wind turbines/wind farms in some countries, but again these simple approaches are very conservative. Large wind farms have been successfully operated on distribution networks with a ratio of fault level to rated capacity as low as 6 with no difficulties.

If system studies are undertaken to investigate the effect of embedded generators on the network voltage then these can either consider the impact on the voltage received by customers or may be based on permissible voltage variations of some intermediate section of the distribution network. Studies considering the effect of embedded generation on, for example, the 11 kV network voltage are simpler to carry out but tend to give more restrictive results than those actually considering the effect on the voltage received by network customers.

Some distribution utilities use more sophisticated control of the on-load tap changers of the distribution transformers including the use of a current signal compounded with the voltage measurement. One technique is that of line drop compensation [8] and, as this relies on an assumed power factor of the load, the introduction of embedded generation and the subsequent change in power factor may lead to incorrect operation if the embedded generator is large compared to the customer load.

Table 1.5   Design rules sometimes used for an indication if an embedded generator may be connected

Network location	Maximum capacity of embedded generator
out on 400 V network	50 kVA
at 400 V busbars	200–250 kVA
out on 11 kV or 11.5 kV network	2–3 MVA
at 11 kV or 11.5 kV busbars	8 MVA
on 15 kV or 20 kV network and at busbars	6.5–10 MVA
on 63 kV to 90 kV network	10–40 MVA

1.5.2 Increase in network fault levels

Most embedded generation plant uses rotating machines and these will contribute to the network fault levels. Both induction and synchronous generators will increase the fault level of the distribution system although their behaviour under sustained fault conditions differs.

In urban areas where the existing fault level approaches the rating of the switchgear, the increase in fault level can be a serious impediment to the development of embedded generation. Uprating of distribution network switchgear can be extremely expensive and, under the charging policies currently used in the UK, this cost will be borne by the embedded generator. The fault level contribution of an embedded generator may be reduced by introducing an impedance between the generator of the network by a transformer or a reactor but at the expense of increased losses and wider voltage variations at the generator. In some countries explosive fuse type fault current limiters are used to limit the fault level contribution of embedded generation plant.

1.5.3 Power quality

Two aspects of power quality are usually considered to be important: (i) transient voltage variations and (ii) harmonic distortion of the network voltage. Depending on the particular circumstance, embedded generation plant can either decrease or increase the quality of the voltage received by other users of the distribution network.

Embedded generation plant can cause transient voltage variations on the network if relatively large current changes during connection and disconnection of the generator are allowed. The magnitude of the current transients can, to a large extent, be limited by careful design of the embedded generation plant, although for single generators connected to weak systems the transient voltage variations caused may be the limitation on their use rather than steady-state voltage rise. Synchronous generators can be connected to the network with negligible disturbance if synchronised correctly, and antiparallel soft-start units can be used to limit the magnetising inrush of induction generators to less than rated current. However, disconnection of the generators when operating at full load may lead to significant, if infrequent, voltage drops. Also, some forms of prime-mover (e.g. fixed speed wind turbines) may cause cyclic variations in the generator output current which can lead to so-called 'flicker' nuisance if not adequately controlled [9,10]. Conversely, however, the addition of embedded generation plant acts to raise the distribution network fault level. Once the generation is connected any disturbances caused by other customers' loads, or even remote faults, will result in smaller voltage variations and hence improved power quality. It is interesting to note that one conventional approach to improving the power quality of sensitive high value manufacturing plants is to install local generation.

Similarly, incorrectly designed or specified embedded generation plants, with power electronic interfaces to the network, may inject harmonic currents which can lead to unacceptable network voltage distortion. However, directly connected generators can also lower the harmonic impedance of the distribution network and so reduce the network harmonic voltage at the expense of increased harmonic currents in the generation plant and possible problems due to harmonic resonances. This is of particular importance if power factor correction capacitors are used to compensate induction generators.

A rather similar effect is shown in the balancing of the voltages of rural MV systems by induction generators. The voltages of rural MV networks are frequently unbalanced due to the connection of single-phase loads. An induction generator has a very low impedance to unbalanced voltages and will tend to draw large unbalanced currents and hence balance the network voltages at the expense of increased currents in the generator and consequent heating.

1.5.4 Protection

A number of different aspects of embedded generator protection can be identified:

- protection of the generation equipment from internal faults
- protection of the faulted distribution network from fault currents supplied by the embedded generator
- anti-islanding or loss-of-mains protection
- impact of embedded generation on existing distribution system protection

Protecting the embedded generator from internal faults is usually fairly straightforward. Fault current flowing from the distribution network is used to detect the fault, and techniques used to protect any large motor are generally adequate. In rural areas, a common problem is ensuring that there will be adequate fault current from the network to ensure rapid operation of the relays or fuses.

Protection of the faulted distribution network from fault current from the embedded generators is often more difficult. Induction generators cannot supply sustained fault current to a three-phase close-up fault and their sustained contribution to asymmetrical faults is limited. Small synchronous generators require sophisticated exciters and field forcing circuits if they are to provide sustained fault current significantly above their full load current. Thus, for some installations it is necessary to rely on the distribution protection to clear any distribution circuit fault and hence isolate the embedded generation plant which is then tripped on over/undervoltage, over/under frequency protection or loss-of-mains protection. This technique of sequential tripping is unusual but necessary given the inability of some embedded generators to provide adequate fault current for more conventional protection schemes.

Loss-of-mains protection is a particular issue in a number of countries, particularly where autoreclose is used on the distribution circuits. For a variety of reasons, both technical and administrative, the prolonged operation of a power island fed from the embedded generator but not connected to the main distribution network is generally considered to be unacceptable. Thus a relay is required which will detect when the embedded generator, and perhaps a surrounding part of the network, has become islanded and will then trip the generator. This relay must work within the dead-time of any autoreclose scheme if out-of-phase reconnection is to be avoided. Although a number of techniques are used, including rate-of-change-of-frequency (ROCOF) and voltage vector shift, these are prone to nuisance tripping if set sensitively to detect islanding rapidly.

The neutral grounding of the generator is a related issue because in a number of countries it is considered unacceptable to operate an ungrounded system and so care is required as to where a neutral connection is obtained and grounded.

The loss-of-mains or islanding problem is illustrated in Figure 1.7. If circuit breaker A opens, perhaps on a transient fault, there may well be insufficient fault current to operate circuit breaker B. In this case the generator may be able to continue to supply the load. If the output of the generator is able to match the real and reactive power demand of the load precisely then there will be no change in either the frequency or voltage of the islanded section of the network. Thus it is very difficult to detect reliably that circuit breaker A has opened using only local measurements at B. In the limit, if there is no current flowing though A (the generator is supplying all the load) then the network conditions at B are unaffected whether A is open or closed. It may also be seen that since the load is being fed through the delta winding of the transformer then there is no neutral earth on that section of the network.

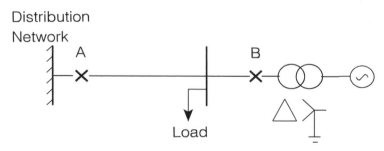

Figure 1.7   Illustration of the 'islanding' problem

Finally, embedded generation may affect the operation of existing distribution networks by providing flows of fault current which were not expected when the protection was originally designed. The fault contribution from an embedded generator can support the network voltage and lead to relays under-reaching.

1.5.5 Stability

For embedded generation schemes, whose object is to generate kWh from new renewable energy sources, considerations of generator transient stability tend not to be of great significance. If a fault occurs somewhere on the distribution network to depress the network voltage and the embedded generator trips, then all that is lost is a short period of generation. The embedded generator will tend to overspeed and trip on its internal protection. The control scheme of the embedded generator will then wait for the network conditions to be restored and restart automatically. Of course if the generation scheme is intended mainly as a provider of steam for a critical process then more care is required to try to ensure that the generator does not trip for remote network faults. However, as the inertia of embedded generation plant is often low and the tripping time of distribution protection long, it may not be possible to ensure stability for all faults on the distribution network.

In contrast, if an embedded generator is viewed as providing support for the power system then its transient stability becomes of considerable importance. Both voltage and/or angle stability may be significant depending on the circumstances. A particular problem in some countries is nuisance tripping of rocof relays. These are set sensitively to detect islanding but, in the event of a major system disturbance (e.g. loss of a large central generator), may mal-operate and trip large amounts of embedded generation. The effect of this is, of course, to depress the system frequency further. The restoration, after an outage, of a section of the distribution network with significant embedded generation may also require care. If the circuit was relying on the embedded generation to support its load then, once the circuit is restored, the load will demand power before the generation can be reconnected. This is, of course, a common problem faced by operators of central generation/transmission networks but is unusual in distribution systems.

Synchronous generators will pole-slip during transient instability but when induction generators overspeed they draw very large reactive currents which then depress the network voltage further and lead to voltage instability. The steady-state stability limit of induction generators can also limit their application on very weak distribution networks because a very high source impedance, or low network short-circuit level, can reduce their peak torque to such an extent that they cannot operate at the rated output.

1.5.6 Network operation

Embedded generation also has important consequences for the operation of the distribution network in that circuits can now be energised from a number of points. This has implications for policies of isolation and earthing for safety before work is undertaken. There may also be more difficulty in obtaining outages for planned maintenance and so reduced flexibility for work on a network with embedded generation connected to it.

1.6 Economic impact of embedded generation on the distribution system

The issues of the technical impact of embedded generation on distribution systems are generally well known and calculation and design techniques well established. The only new issues emerging are those concerned with the overall penetration magnitude of the embedded generation and the particular difficulties of very large schemes being connected to very weak networks. In contrast, the economic impact of embedded generation on distribution systems is only now being seriously addressed.

Embedded generation alters the power flows in the network and so will alter network losses. If a small embedded generator is located close to a large load then the network losses will be reduced as both real and reactive power can be supplied to the load from the adjacent generator. Conversely, if a large embedded generator is located far away from network loads then it is likely to increase losses on the distribution system. A further complication arises due to differing values of electrical energy as the network load increases. In general there is a correlation between high load on the distribution network and the use of expensive generation plant. Thus, any embedded generator which can operate in this period and reduce distribution network losses will make a significant impact on the costs of operating the network.

At present, embedded generation generally takes no part in the voltage control of distribution networks. Thus, in the UK, embedded generators will generally choose to operate at unity power factor to minimise their electrical losses and avoid any charges for reactive power consumption, irrespective of the needs of the distribution network. In Denmark some progress has been made in this regard with embedded CHP schemes operating at three different power factors according to the time of day [11]. During periods of peak loads reactive power is exported to the network while during low network load the generators operate at unity power factor.

Embedded generation can also, in principal, be used as a substitute for distribution network capacity. Clearly, embedded generators cannot substitute for radial feeders, as islanded operation is not generally acceptable, and network extensions may be required to collect power from isolated renewable energy schemes. However, most high-voltage distribution circuits are duplicated or meshed, and embedded generation can reduce the requirement for these assets. At present the concept, that embedded generation can substitute for distribution circuit capacity, is not generally accepted by distribution utilities.

1.7 Impact of embedded generation on the transmission system

In a similar manner to the distribution system, embedded generation will alter the flows in the transmission system. Hence transmission losses will be altered, generally reduced, while in a highly meshed transmission network it is easier to demonstrate that reduced flows lead to a lower requirement for assets. In the UK, the charges for use of the transmission network are currently evaluated based on a measurement of peak demand at the Grid Supply Point. When embedded generation plant can be shown to be operating during the periods of peak demand then it is clearly reducing the charges for use of the transmission network.

1.8 Impact of embedded generation on central generation

The main impact of embedded generation on central generation is to reduce the mean of the power output of the central generators but, often, to increase the variance. In a large electrical power system, consumer demand can be estimated quite accurately by the generator despatching authority. Embedded generation will introduce additional uncertainty in these estimates and so may require additional reserve plant. In Denmark [11,12], considerable effort has been made to predict the output of wind farms by forecasting wind speeds, and embedded CHP plants by forecasting heat demand. Both forecasts are based on meteorological techniques. As embedded generation supplies an increasing proportion of the customer load, particularly during times of low demand, the provision of generation reserve and frequency control becomes an important issue. Conventional central generating plant (i.e. steam or hydro-sets) is able to provide these important ancillary services which are necessary for the power system to function. If embedded generation displaces such plant then these services must be provided by others and the associated additional costs will then reduce the value of the embedded generation output. This point is discussed in detail in the evidence of the National Grid Company to the House of Lords Select Committee [7].

1.9 References

1 CIGRE STUDY COMMITTEE No. 37: 'Impact of increasing contributions of dispersed generation on the power systems'. Final report of Working Group 37–23, September 1998. To be published in *Electra*
2 CIRED preliminary report of CIRED Working Group 04: 'Dispersed generation'. Issued at the CIRED Conference in Nice, June 1999
3 SCHWEER, A.: 'Special report for Session 3'. Paper no. 300–00, CIGRE Symposium on *Impact of demand side management, integrated resource planning and distributed generation*, Neptun, Romania, 17–19 September 1997
4 PETRELLA, A.J.: 'Issues, impacts and strategies for distributed generation challenged power systems'. Paper no. 300–12, CIGRE Symposium on *Impact of demand side management, integrated resource planning and distributed generation*, Neptun, Romania, 17–19 September 1997
5 EnergieNed: 'Energy distribution in the Netherlands, 1997.'
6 COMMISSION OF THE EUROPEAN UNION: 'Energy for the future: renewable sources of energy'. White paper for a community strategy and action plan, COM 97, 599.
7 HOUSE OF LORDS SELECT COMMITTEE ON THE EUROPEAN COMMUNITIES: 'Electricity from renewables'. HL Paper 78–1, June 1999, The Stationery Office
8 LAKERVI, E., and HOLMES, E.J.: 'Electricity distribution network design' (Peter Peregrinus, London, 1989)
9 HEIER, S.: 'Grid integration of wind energy conversion systems' (John Wiley and Sons, Chichester, 1998)
10 DUGAN, S., MCGRANAGHAN, M.F., and BEATY, H.W.: 'Electrical power systems quality' (McGraw Hill, New York, 1996)
11 JORGENSEN, P., GRUELUND SORENSEN, A., FALCK CHRISTENSEN, J., and HERAGER, P., 'Dispersed CHP units in the Danish Power System'. Paper no 300–11, CIGRE Symposium on *Impact of demand side management, integrated resource planning and distributed generation*, Neptun, Romania, 17–19 September 1997
12 FALCK CHRISTENSEN, J., PEDERSEN, S., PARBO, H., GRUELUND SORENSEN, A., and TOFTING, J., 'Wind power in the Danish Power System'. Paper no. 300–09, CIGRE Symposium on *Impact of demand side management, integrated resource planning and distributed generation*, Neptun, Romania, 17–19 September 1997

Chapter 2

Embedded generation plant

There are many varied types of generating plant connected to electrical distribution networks ranging from well established equipment such as Combined Heat and Power (CHP) units and internal combustion reciprocating engines to more recent types of generation such as wind farms and photovoltaics. In the future some of the many emerging technologies, such as fuel cells, flywheel storage, micro-CHP using small gas turbines or Stirling engines, may become commercially significant. Understanding the interaction of embedded generation with the power system requires an appreciation of the technology of the prime movers, the characteristics of the energy sources and also the conditions under which the embedded generation plant is operated. In deregulated electricity supply systems it is also important to recognise that the owners of embedded generation plant (who will not be the distribution utility in many cases) will respond to pricing and other commercial signals to determine whether to invest in such plant and then how to operate it. This chapter is intended to provide a brief introduction to several of the important embedded generation technologies and to emphasise the multidisciplinary aspects of embedded generation.

2.1 Combined Heat and Power plants

Combined Heat and Power (CHP) is, at present, the most significant type of generation embedded in distribution systems. CHP, sometimes known as cogeneration, is the simultaneous production of electrical power and useful heat. Generally the electrical power is consumed inside the host premises or plant of the CHP facility, although any surplus or deficit is exchanged with the utility distribution system. The heat generated is either used for industrial processes and/or for space heating inside the host premises or alternatively is transported to the local area for district heating. Industrial CHP schemes typically achieve a 35% reduction in primary energy use compared with electrical generation from central power stations and heat-only boilers. In the UK power system, this leads to a reduction in CO_2 emissions of over 30% in comparison with large coal fired stations and over 10% in comparison with central combined-cycle gas turbine plant [1].

The use of CHP for district heating is limited in the UK, although small schemes exist in some cities. However, in northern Europe, e.g. Denmark, Sweden and Finland, district heating is common in many large towns and cities, with the heat supplied at water temperatures in the range 80–150 °C, either from CHP plant or heat-only boilers [2]. A recent development in Denmark has been to extend the use of CHP into rural areas with the installation of smaller CHP schemes in villages and small towns using either back-pressure steam turbines, fed from biomass in some cases, or reciprocating engines powered by gas [3].

Tables 2.1, 2.2 and 2.3 summarise the position of CHP in the UK in 1998 [1]. The total capacity was 3929 MWe, which may be compared with the government policy target of 5000 MWe by the year 2000 and 10 000

Table 2.1   Summary of CHP installations in the UK in 1998 [1]

Electrical capacity, MWe	3 929
Heat capacity, MWth	15 500
Overall efficiency, %	69

Table 2.2   CHP installations by size in the UK in 1998 [1]

Size range	Number of installations	Total electricity capacity, MWe
<100 kWe	674	37.8
100–999 kWe	469	119.6
1–9.9 MWe	161	664.2
>10 MWe	72	3 107.0
Total	1 376	3 928.6

Table 2.3   CHP equipment installed in the UK in 1998 [1]

Main prime mover	Electrical capacity, MWe	Average heat/power ratio
Back-pressure steam turbine	1 300	6.9:1
Pass-out condensing steam turbine	54	6.7:1
Gas turbine	839	3.6:1
Combined cycle	1 283	1.8:1
Reciprocating engine	453	1.8:1

MWe by 2010. The high overall efficiency of nearly 70% confirms the potential of CHP in reducing environmental impact as well as for financial savings: 321 industrial sites make up 88% of the CHP capacity, while the remaining 1055 sites are in the commercial, public and residential sectors. Table 2.2 shows that the installed capacity is dominated by a small number of large installations while the largest number of sites have an electrical capacity of less than 100 kWe. Some of the very large CHP installations, e.g. on chemical works and oil refineries, are connected electrically to the transmission network, and so this generation cannot be considered to be embedded. In contrast, many of the CHP schemes installed in buildings in the leisure, hotel and health sectors have installed electrical capacities below 1 MWe and are connected to the 11 kV distribution network.

Table 2.3 refers to the various technologies used in CHP plants. Back-pressure steam turbines exhaust steam at greater than atmospheric pressure either directly to an industrial process or to a heat exchanger. The higher the back pressure the more energy there is in the exhausted steam and so the less electrical power that is produced. The back-pressure steam turbines in the UK have an average heat to power ratio of 7:1 and so, once the site electrical load has been met, any export of electrical power will be small. Figure 2.1 is a simplified diagram of a CHP scheme using a back-pressure steam turbine. All the steam passes through the turbine which drives a synchronous generator, usually operating at 3000 rpm. After the turbine, the steam, at a pressure typically in the range 0.12–4 MPa and a temperature of between 200 and 300 °C [4] depending on its use, is passed to the industrial process or through a heat exchanger for use in space heating.

In contrast, in a pass-out (or extraction) condensing steam turbine (Figure 2.2) some steam is extracted at an intermediate pressure for the

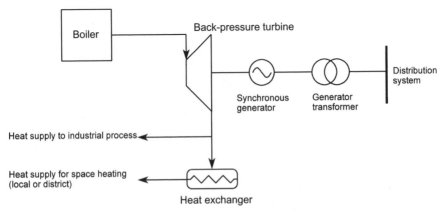

Figure 2.1   CHP scheme(s) using a back-pressure steam turbine

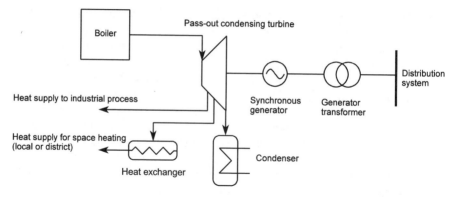

Figure 2.2   CHP scheme(s) using a pass-out condensing steam turbine

supply of useful heat with the remainder being fully condensed. This arrangement allows a wide range of heat/power ratios. In Denmark all the large town and city district heating schemes (150–350 MWe) use this type of unit which, taking into account the capacity which is reduced for heat load, can be despatched in response to the public utilities' electrical power demand.

Figure 2.3 shows how the waste heat from a gas turbine may be used. Gas turbines using either natural gas or distillate oil liquid fuel are available in ratings from less than 1 MWe to more than 100 MWe, although at the lower ratings internal combustion reciprocating engines may be preferred. In some industrial installations, provision is made for supplementary firing of the waste heat boiler to ensure the availability of useful heat when the gas turbine is not operating or to increase the heat/power ratio. The exhaust gas temperature of a gas turbine can be as high as 500–600 °C and so, with the large units, there is the potential to increase the generation of electrical power by adding a steam turbine and to create a combined-cycle plant. Steam is raised using the exhaust gases of the gas turbine passed through a waste heat boiler which is fed to either a back-pressure or pass-out condensing steam turbine. Useful heat is then recovered from the steam turbine. In CHP schemes using a combined cycle some 80–90% of the energy in the fuel is transformed into electrical power or useful heat, with a lower heat/power ratio than that of single-cycle gas turbines (see Table 2.3). Combined-cycle CHP plants, because of their complexity and capital cost, tend to be suitable for large electric and heating loads such as the integrated energy supply to a town or large industrial plant [5].

Table 2.2 shows that, although in the UK installed CHP capacity is dominated by plants with ratings greater than 10 MWe, there are currently over 1000 CHP installations with ratings less than 1 MWe, and over 600 with ratings less than 100 kWe. These plants of less than 100

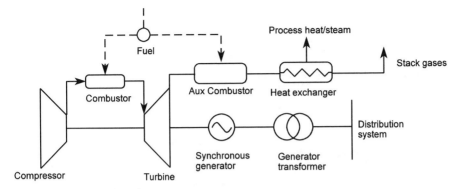

Figure 2.3   CHP scheme using a gas turbine with waste heat recovery

kWe are typically skid mounted units consisting of a reciprocating 4-stroke engine driving a three-phase synchronous, or in some cases asynchronous, generator with a heat recovery system to extract heat from the exhaust gases, the cooling water, and the lubrication oil [6] (see Figure 2.4). For larger engines (i.e. approaching 500 kW) it becomes economic to pass the exhaust gas, which may be at up to 350–400 °C, to a steam-producing waste heat boiler. The heat available from the cooling jacket and the lubrication oil is typically at 70–80 °C. The fuel used is usually natural gas, sometimes with a small addition of fuel oil to aid combustion, or, in some cases, digester gas from sewage treatment plants. Typical applications include leisure centres, hotels, hospitals, academic establishments and industrial processes. Landfill gas sites tend to be too far away from a suitable heat load and so the engines, fuelled by the landfill gas, are operated as electrical generating sets only, and not in CHP mode. The economic case for a CHP scheme depends both on the heat and electrical power load of the host site and critically on the costs of alternative energy supplies and also the rate received for exporting electrical power to the utility. Reference 6 lists the principal economic attractions of small-scale CHP in the UK in 1991 as being due to savings from: (i) the high cost of electrical energy from the utility, (ii) high electrical maximum demand charges and (iii) reduction in inefficient boiler operation.

CHP units are typically controlled, or despatched, to meet the energy needs of the host site and not to export electrical power to the utility distribution system [7]. It is common for CHP units to be controlled to meet a heat load and, in district heating schemes, the heat output is often controlled as a function of ambient temperature. Alternatively, the units can be controlled to meet the electrical load of the host site and any deficit in the heat requirement is met from an auxiliary source. Finally, the units may be run to supply both heat and electricity to the site in an

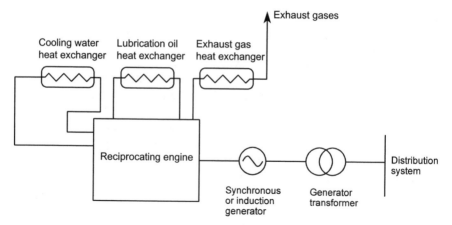

Figure 2.4   CHP scheme using a reciprocating engine with heat recovery

optimal manner, but this is likely to require a more sophisticated control system.

Although CHP schemes are conventionally designed and operated to meet the energy needs of the host site, or a district heating load, this is a commercial/economic choice rather than being due to any fundamental limitation of the technology. As commercial and administrative conditions change, perhaps in response to the policy drivers described in Chapter 1, it may be that CHP plants will start to play a more active role in supplying electrical energy and other ancillary services to the electricity distribution system. For example, the main objective of the Danish dispersed CHP plants described in Reference 3 is to provide 95% of the annual requirement for district heating within their area of operation. The direct relationship between heat and electrical power production from these mainly reciprocating engine units was a major concern as it would have imposed significant additional load variations on the larger electrical power generating units when the dispersed CHP units responded to the varying demand for district heating. Therefore, large heat stores were constructed for each district heating scheme to accommodate approximately 10h of maximum heat production. One benefit of the heat stores is that they allow the CHP units to be run for reduced periods but at rated output, and hence maximum efficiency. Also, the periods of operation can be chosen to be at periods of maximum demand on the electrical power network. A three-rate tariff was introduced for the electrical power generated by the CHP units. The high kWh tariff corresponds to the demand for electricity on the utility system and, together with the facility to store heat, ensures that the CHP units generate electrical power when it is most needed by, and hence is most valuable to, the electricity system.

2.2 Renewable energy generation

The location of embedded generators used for CHP schemes is fixed by the position of the heat load, and their operation is generally controlled in response to the energy demands of the host site or of a district heating scheme. Similarly, the siting of generators using renewable energy sources is determined by the location of the renewable energy resource, and their output follows the availability of the resource. Certainly, the choice of location of any generation plant is limited by the local environmental impact it may have, but it is obvious that, for example, the position of a small-scale hydro-generating station (and hence the possibilities for connecting it to the distribution network) must be determined by the location of the hydro-resource. Unless the renewable energy resource can be stored (e.g. as potential energy in dams or by storing biomass), the generator must operate when the energy is available. In general it is not cost-effective to provide large energy stores for small renewable energy schemes, and so their output varies with the available resource. This is an important difference compared with embedded generators that use fossil fuel which, because of its much higher energy density, can be stored economically.

Table 2.4 shows the capacity and electricity generated from renewables in the UK at the end of 1998, while Table 2.5 shows the contracts awarded by the UK Government under the 5th round of the Non Fossil Fuel Obligation (NFFO) in September 1998. The NFFO was a competitive arrangement in England and Wales to encourage the establishment of renewable energy generation. Similar arrangements apply to Scotland

Table 2.4   Capacity of and electricity generated from renewables in 1998 [1]

Technology	Declared net capacity, MWe	Generation, GWh	Load factor, %
Onshore wind	139.5	886	31
Small-scale hydro	59.0	204 ⎫	37.6
Large-scale hydro	1 413.0	5 022 ⎭	
Biofuels:			
landfill gas	224.5	1 180	
sewage sludge digestion	83.0	386	
municipal solid waste combustion	165.1	1 351	
Other biofuels	84.2	318	
Total biofuels	556.8	3 235	65.5

Declared net capacity of onshore wind is 0.43 of installed capacity
Load factor is based on installed capacity rather than on declared net capacity

Table 2.5   Size and composition of the support for generation from new renewable energy sources under NFFO-5, September 1998

Technology	Number of projects contracted	Capacity of projects contracted, MW	Average contract price, p/kWh
Landfill gas	141	14	2.73
Energy from waste	22	416	2.43
Energy from waste using CHP	7	70	2.63
Small-scale hydro	22	9	4.08
Wind energy exceeding 0.995 MW	33	340	2.88
Wind energy not exceeding 0.995 MW	36	28	4.18
Total	261	1 177	2.71

Capacity of projects is declared net capacity

and Northern Ireland. Potential developers of renewable energy generating schemes bid the price, in p/kWh, at which they require to construct their schemes, and contracts are then awarded on the basis of this price. Under the NFFO-5 order the price is guaranteed for a period of 15 years with an annual increase related to inflation. The energy is bought by the local distribution company at the point of connection of the embedded generation plant with the distribution network. Other European countries have adopted different approaches to the support of renewable energy generation, including the guarantee of a certain percentage of the retail electricity price (e.g. the German 'Electricity Feed Law') or more direct government intervention. From 2000 the NFFO scheme will be replaced by a support mechanism based on a system of 'Green Certificates' and an obligation on electricity suppliers to provide a minimum percentage of energy from renewable sources.

Table 2.5 shows that, under NFF0–5, 261 schemes with a capacity of 1177 MW were awarded contracts, bringing the total capacity awarded (including the preceding renewable obligations throughout the UK) to some 880 schemes with a total rating of 3493 MW [1]. However, difficulties in developing the projects, including problems with obtaining the necessary planning consents, will result in only some fraction of it being commissioned. The name-plate rating of the embedded generation plant is higher than that given in Table 2.5 as the capacity is given in declared net capacity (DNC). This is the name-plate capacity reduced by the internal consumption of the generation plant and with a factor applied to take account of the intermittent nature of the resource. For wind energy the factor relating DNC to name-plate capacity is 0.43 and so the name-plate capacity of wind energy (exceeding 0.995 MW) is 340/0.43 or 791 MW.

Table 2.5 indicates that small-scale hydro is, at present, one of the more expensive of the technologies, reflecting the scarcity of good new small hydro-sites in England and Wales. Burning landfill gas in internal combustion engines is clearly attractive. The average capacity of these projects is just over 2 MW and so these units, which use synchronous generators, would be connected to MV distribution networks without difficulty. The energy from waste schemes, predominantly combustion of municipal or industrial waste, is large, with a mean capacity of almost 20 MW. Again these will use conventional synchronous generators although connection at 33 or 132 kV is likely to be required. Wind energy is treated in two bands to reflect the importance of fixed project costs irrespective of the capacity of the plant: 33 of the larger wind farms received contracts with a mean name-plate capacity of 24 MW. These require particular attention because of: (i) their size (the largest scheme was 97 MW); (ii) their likely remote locations in areas of good wind resource but away from strong electrical networks; (iii) their use of induction generators; (iv) the variable nature of their output.

2.2.1 Small-scale hydro-generation

The operation of small and medium sized hydro-generating units in parallel with the distribution system is now well understood [8,9]. However, those hydro-schemes without significant storage capacity may experience very large variations in available water flow, and hence output, particularly if the catchment (the area of land over which the water is gathered) is on rocky or shallow soil with a lack of vegetation cover, and steep, short streams. Clearly uneven rainfall will lead to a variable resource, and it is interesting to note that the capacity factor for hydro in the UK for 1998 was only some 30%. (Capacity factor is the ratio of annual energy generated to that which would be generated with the plant operating at the rated output all year.)

Figure 2.5 shows the average daily flows of two rivers with catchment areas of very different characteristics leading to very different hydrological resources. It is conventional to express the resource as a flow duration curve (FDC), which shows the percentage time that a given flow is equalled or exceeded (Figure 2.6) [10]. Although an FDC gives useful information about the annual energy yield which may be expected from a given hydro-resource it provides no information as to how the output of the hydro-scheme might correlate with the day-to-day load demand on the power system. The power output of a hydro-turbine is given by the simple expression

$$P = QH\eta\rho g \tag{2.1}$$

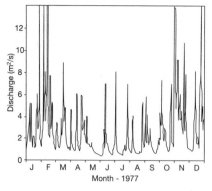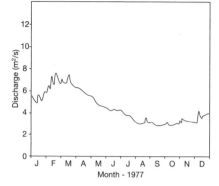

Figure 2.5   Average daily flows of two rivers

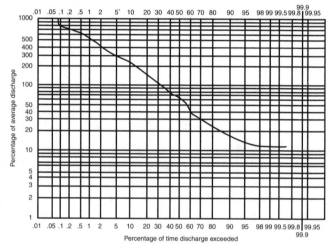

Figure 2.6   Typical flow duration curve

where:

P = output power (W)
Q = flow rate (m³/s)
H = effective head (m)
η = overall efficiency
ρ = density of water (1000 kg/m³)
g = acceleration due to gravity.

Clearly a high effective head is desirable as this allows increased power output while limiting the required flow rate and hence cross-section of penstock (the pressurised pipe bringing water to the turbine). Various forms of turbine are used for differing combinations of flow rate and head [11]. At lower heads, reaction turbines operate by changing the direction of the flow of water. The water arriving at the runner is still under pressure and the pressure drop across the turbine accounts for a significant part of the energy extracted. Common designs of reaction turbine include Francis and propeller (Kaplan) types. At higher heads, impulse turbines are used, e.g. Pelton or Turgo turbines. These operate by extracting the kinetic energy from a jet of water which is at atmospheric pressure. For small hydro-units (<100 kW) a cross-flow impulse turbine, where the water strikes the runner as a sheet rather than a jet, may be used. Typically, reasonable efficiencies can be obtained with impulse turbines down to 1/6th of rated flow, whereas, for reaction turbines efficiencies are poor below 1/3 of rated flow. This has obvious implications for the rating of plant given the variable resource of some catchments.

Small-scale hydro-schemes may use induction or synchronous generators. Low head turbines tend to run more slowly and so either a gearbox or a multipole generator is required. One particular design consideration is to ensure that the turbine-generator will not be damaged in the event of the electrical connection to the network being broken and the hence the load being lost and the turbine-generator overspeeding [9].

2.2.2 *Wind power plants*

A wind turbine operates by extracting kinetic energy from the wind passing through its rotor. The power developed by a wind turbine is given by

$$P = \tfrac{1}{2} \, C_p \rho V^3 A \tag{2.2}$$

where:

C_p = power coefficient
P = power (W)
V = wind velocity (m/s)
A = swept area of rotor disc (m^2)
ρ = density of air (1.225 kg/m^3)

As the power developed is proportional to the cube of the wind speed it is obviously important to locate any electricity generating turbines in areas of high mean annual wind speed, and the available wind resource is an important factor in determining where wind farms are sited. Often the areas of high wind speed will be away from habitation and the associated well developed electrical distribution network, leading to a requirement for careful consideration of the integration of wind turbines to relatively weak electrical distribution networks. The difference in the density of the working fluids (water and air) illustrates clearly why a wind turbine rotor of a given rating is so much larger than a hydro-turbine. A 1.5 MW wind turbine will have a rotor diameter of some 60 m mounted on a 60–90 m high tower. The force exerted on the rotor is proportional to the square of the wind speed and so the wind turbine must be designed to withstand large forces during storms. Most modern designs use a three-bladed horizontal-axis rotor as this gives a good value of peak C_p together with an aesthetically pleasing design.

The power coefficient C_p is a measure of how much of the energy in the wind is extracted by the turbine rotor. It varies with rotor design and the relative speed of the rotor and wind (known as the tip speed ratio) to give a maximum practical value of approximately 0.4 [12].

Figure 2.7 is the power curve of a wind turbine which indicates its output at various wind speeds. At wind speeds below cut-in (5 m/s) no significant power is developed. The output power then increases rapidly with wind speed until it reaches its rated value and is then limited by some control action of the turbine. This part of the characteristic follows an approximately cubic relationship between wind speed and output power, although this is modified by changes in C_p. Then at the shut-down wind speed (25 m/s in this case) the rotor is parked for safety.

Figure 2.8 shows a typical annual distribution of hourly mean wind speeds from a UK lowland site and, by comparison with Figure 2.7, it may be seen that the turbine will only be operating at the rated output for some 10–15% of the year. Depending on the site wind speed distribution the turbine may be shut down due to low winds for up to 25 % of the year, and during the remaining period the output will fluctuate with wind speed. Figures 2.7 and 2.8 can be used to calculate the annual energy yield from a wind turbine but they provide no information as to when the

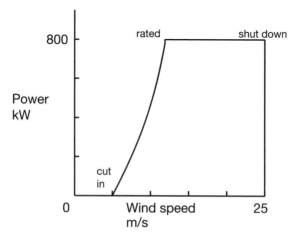

Figure 2.7   Wind turbine power curve

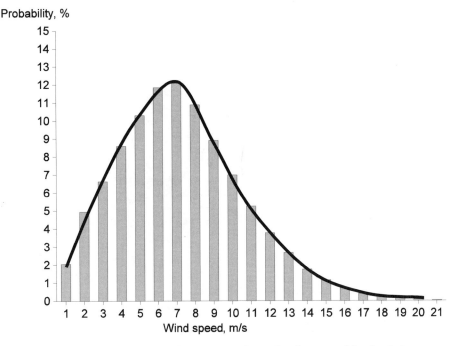

Figure 2.8   Distribution of hourly mean wind speeds of a typical lowland site

energy is generated. Figure 2.9 shows a time series of the power outputs of a wind farm in the UK, while Table 2.6 shows the capacity factors of a number of wind farms by season.

The wind speed distribution of Figure 2.8 is formed from hourly mean wind speeds and the power curve is measured using 10 min average data. There are also important higher-frequency effects which, although they do not significantly influence the energy generated, are important for their consequences on the machinery and on network power quality. With some designs of fixed-speed wind turbines, it has been found to be rather difficult to limit the output power to the rated value indicated by the horizontal section of the power curve (Figure 2.7), and transient overpowers of up to twice nominal output have been recorded in some instances. Therefore, when considering the impact of the turbines on the distribution network, it is important to establish what the maximum transient output power is likely to be.

The torque from a horizontal axis wind turbine rotor contains a peri-odic component at the frequency at which the blades pass the tower [13]. This cyclic torque is due to the variation in wind speed seen by the blade as it rotates. The variation in wind speed is due to a combination of: tower shadow, wind shear and turbulence. In a fixed speed wind turbine, this rotor torque variation is then translated into a change in the output power and hence a voltage variation on the network. These dynamic

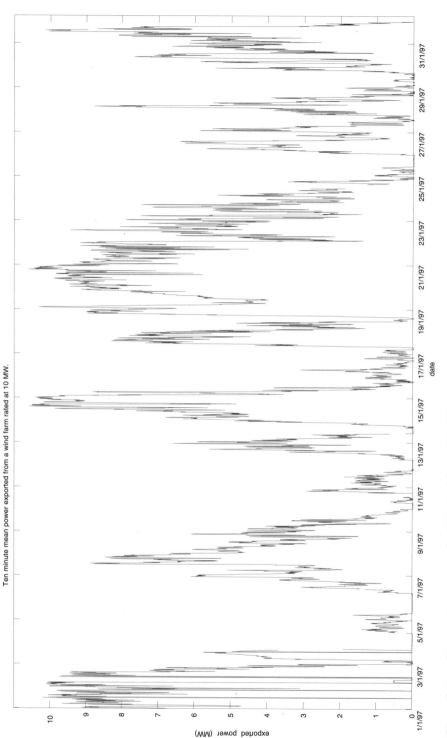

Figure 2.9a Time series output of a UK wind farm (Real Power)
Data courtesy of National Wind Power

Ten minute mean reactive power imported from a wind farm rated at 10 MW.

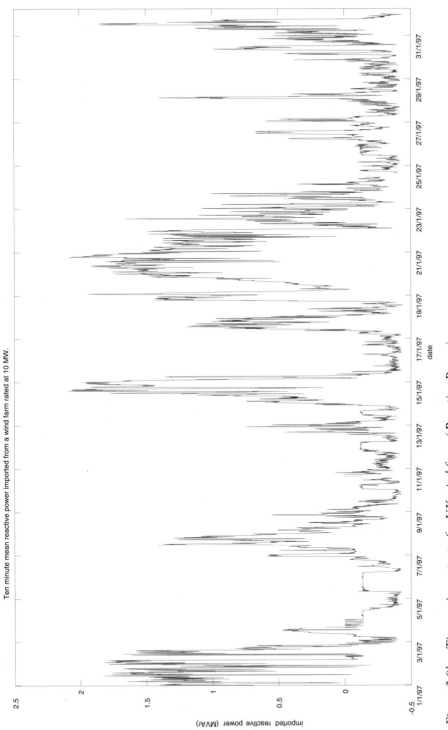

Figure 2.9b   Time series output of a UK wind farm (Reactive Power)
Data courtesy of National Wind Power

Table 2.6   Monthly capacity factors of typical UK wind farms

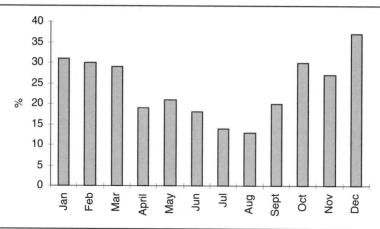

voltage variations are often referred to as 'flicker' because of the effect they have on incandescent lights. The human eye is very sensitive to changes in light intensity, particularly if the variation occurs at frequencies around 10 Hz. For a large wind turbine the blade passing frequency will be around 1–2 Hz and, although the eye is less sensitive at this frequency, it will still detect voltage variations greater than about 0.5%. In general the torque fluctuations of individual wind turbines in a wind farm are not synchronised and so the effect in large wind farms is reduced as the variations average out. The impact of wind turbines, and other embedded generators, on network power quality is discussed in more detail in Chapter 5.

Although reliable commercial wind turbines can be bought from a variety of manufacturers there is still very considerable development of the technology, particularly as the size and ratings of turbines increase. Some of the major differences in design philosophy include: (i) fixed- or variable-speed operation, (ii) direct drive generators or the use of a gearbox and (iii) stall or pitch regulation.

Fixed-speed wind turbines using induction generators are simpler and, it may be argued, more robust. It is not usual to use synchronous generators on network-connected fixed-speed wind turbines as it is not practicable to include adequate damping in a synchronous generator rotor to control the periodic torque fluctuations of the aerodynamic rotor. Some very early wind turbine designs did use synchronous generators by including mechanical damping in the drive trains (e.g. by using a fluid coupling) but this is no longer common practice. Figure 2.10 shows a simplified schematic diagram of a fixed-speed wind turbine. The aerodynamic rotor is coupled to the induction generator via a speed-increasing gearbox. The induction generator is typically wound for

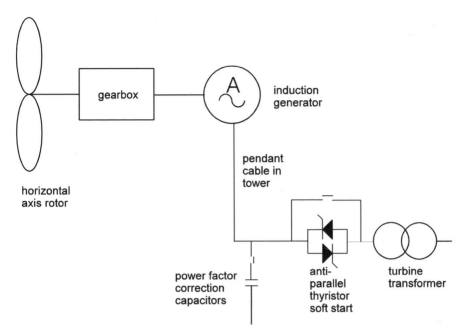

Figure 2.10   Schematic diagram of fixed speed wind turbine

690 V, 1000 or 1500 rpm operation. Pendant cables within the tower connect the generator to switched power factor correction capacitors and an antiparallel soft-start unit located in the tower base. It is common to bypass the soft-start thyristors once the generator has been fully fluxed. The power factor correction capacitors are either all applied as soon as the generator is connected or they are switched in progressively as the average output power of the wind turbine increases. It is not conventional to control these capacitors based on network voltage. A local transformer, typically 690V/33 kV in UK wind farms, is then located either inside the tower or adjacent to it.

With variable-speed operation it is possible, in principle, to increase the energy captured by the aerodynamic rotor by maintaining the optimum power coefficient over a wide range of wind speeds. However, it is then necessary to decouple the speed of the rotor from the frequency of the network through some form of power electronic converter. If a wide range of variable-speed operations is required then the arrangement shown in Figure 2.11 may be used. Two voltage source converter bridges are used to interface the wind turbine drive train to the network. The generator side bridge is commonly used to maintain the voltage of the DC link, while the network side converter is then used to control the output power and hence torque on the rotor. The generator may be either synchronous or induction, and some form of pulse width modulated

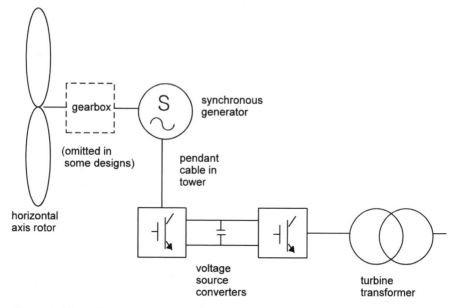

Figure 2.11   Schematic diagram of variable speed wind turbine

(PWM) switching pattern is used on the converter bridges. As all the power is transferred through the variable-speed equipment, there are significant electrical losses and, at low output powers, the energy gains from variable-speed operation of the rotor may not be realised. However, variable-speed operation gives a number of important other advantages which follow from the variable speed rotor acting as a flywheel-type energy store. These include: (i) a reduction in mechanical loads which allows lighter mechanical design and (ii) smoother output power. Some designs operate over a restricted speed range (e.g. by using slip energy recovery on a wound rotor induction machine) to gain most of the benefits of variable-speed operation but with lower losses.

Some manufacturers have dispensed with the gearbox between the rotor and the generator and have developed large-diameter direct-drive generators which rotate at the same speed as the aerodynamic rotor. It is not practical to use large-diameter induction generators because the required air-gap is too small, and so either wound rotor or permanent magnet synchronous generators are used. These multipole generators are then interfaced to the network as shown in Figure 2.11.

At the rated wind speed it is necessary to limit the power into the wind turbine rotor and so some form of rotor regulation is required [12]. Stall regulation is a passive system with no moving parts which operates by the rotor blades entering aerodynamic stall once the wind speed exceeds the rated value. It relies on the rotational speed of the rotor being controlled and so is usually found on fixed-speed wind turbines. Pitch regulated rotors have an actuator and control system to rotate the blades about their axes and so limit power by reducing the angle of attack seen by the aerofoil. Pitch regulation requires a more complex control system but can lead to rather higher energy capture. Recently, a combination of these two approaches has been introduced, known as active-stall, where the blade is rotated about its axis but the main power limitation remains aerodynamic stall.

These differences in design philosophy do not fundamentally affect the basic energy production function of a wind turbine, which is defined by Figures 2.7 and 2.8, but they do have an important influence on the dynamic operation of the turbine and the output power quality.

2.2.3 Offshore wind energy

The next major development in wind energy is likely to be major installations offshore. There are already three small 'offshore' wind farms located in rather shallow waters off Denmark and Holland. However, the new wind farms now being considered will be large, typically 50–100 MW in the first instance, and may be located many kilometres offshore.

The advantages of offshore installations include:

- reduced visual impact
- higher mean wind speed
- reduced wind turbulence
- low wind shear leading to lower towers.

The disadvantages include:

- higher capital costs
- access restrictions in poor weather
- submarine cables required.

Initially, the turbines used offshore will be similar to those used on land. However, as the production volumes increase, it is likely that the designs will become tailored for offshore with greater emphasis on reliability and also perhaps increased rotor tip speeds.

The integration of offshore wind farms with distribution networks will pose a number of new challenges, mainly due to their size and remote location. It is also likely that, due to the more uniform wind conditions offshore, the variability in output of a single large offshore wind farm will be greater than an equivalent number of wind turbines distributed more widely on land. Initially the design of the electrical infrastructure of offshore wind farms will be rather similar to that for large land-based installations. However, some potential future offshore sites are in shallow water, or on sand banks, a long way from the coast. Hence serious consideration is also being given to using voltage source HVDC transmission to bring the power ashore and so avoid the problems and expense associated with long AC high-voltage submarine cables.

2.2.4 Solar photovoltaic generation

Photovoltaic generation, or the direct conversion of sunlight to electricity, is a well established technology for power supplies to sites remote from the distribution network, with a number of major manufacturers producing equipment. However, it is now also being seriously considered as a potentially cost-effective means of generating bulk supplies of electricity for general use. Although a number of large (MW scale) demonstration projects have been constructed in the past, interest is now focused on incorporating the photovoltaic modules into the fabric of buildings to reduce overall cost and space requirements. Thus, these rather small PV installations would be connected directly into customers' circuits and so interface with the LV distribution network. This form of generation would be truly embedded with, perhaps, very large numbers of residences and commercial buildings being equipped with photovoltaic generators.

Outside the Earth's atmosphere, the power density of the solar radiation, on a plane perpendicular to the direction to the sun, is approximately 1350 Wm^{-2}. As the sunlight passes through the Earth's atmosphere the energy in certain wavelengths is selectively absorbed. Thus both the spectrum and power density change as the light passes through the atmosphere.

A term known as the air mass (AM) has been introduced to describe the ratio of the path length of the solar radiation through the atmosphere to its minimum value, which occurs when the sun is directly overhead. For the standard testing of photovoltaic modules it is conventional to use a spectrum corresponding to an air mass of 1.5 (AM 1.5) but with the power density adjusted to 1000 Wm^{-2}. A cell temperature of 25°C is also assumed. These are the conditions under which the output of a photovoltaic module is specified and may differ significantly from those experienced in service.

The total or global solar energy arriving at a photovoltaic module comprises direct and diffuse components. Direct radiation consists of the almost parallel rays that come from the solar disc. Diffuse radiation is that scattered in the atmosphere and approaches the module from all parts of the sky. On a clear day the direct component may make up 80–90% of the total radiation, but on a completely overcast day this drops to almost zero leaving a small diffuse component, say 10–20%, of the radiation expected on a clear day. Figure 2.12 shows typical daily curves of

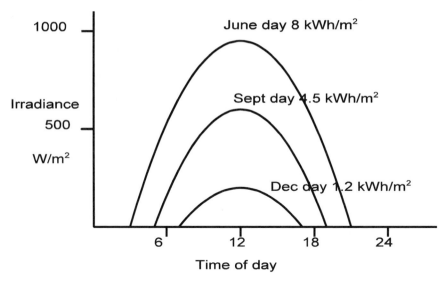

Figure 2.12   Solar irradiance on a flat plane at 48° north

global radiation on a horizontal surface for clear conditions at latitude 48° and illustrates the difference in radiation in the winter and summer months. Mainland England and Scotland extends from latitude 50° to 59° and so the seasonal variations in the solar resource will be even more pronounced in the UK. Figure 2.13 shows the irradiance on a photovoltaic installation in a day of varying cloud conditions, while Figure 2.14 shows the output of a small photovoltaic system at the same site over a week.

A detailed description of the physics of photovoltaic energy conversion is well outside the scope of this book but a useful introduction is given by a number of authors [14–16]. Nevertheless an initial understanding of the performance of a solar cell may be obtained by considering it as a diode in which the light energy, in the form of photons with the appropriate energy level, falls on the cell and generates electron–hole pairs. The electrons and holes are separated by the electric field established at the junction of the diode and are then driven around an external circuit by this junction potential. There are losses associated with the series and shunt resistance of the cell as well as leakage of some of the current back across the *p-n* junction. This leads to the equivalent circuit of Figure 2.15 and the operating characteristic of Figure 2.16. Note that Figure 2.16 is drawn to show the comparison with a conventional diode characteristic. The current produced by a solar cell is proportional to its surface area and the incident irradiance, while the voltage is limited by the forward potential drop across the *p-n* junction.

To produce higher voltages and currents the cells are arranged in

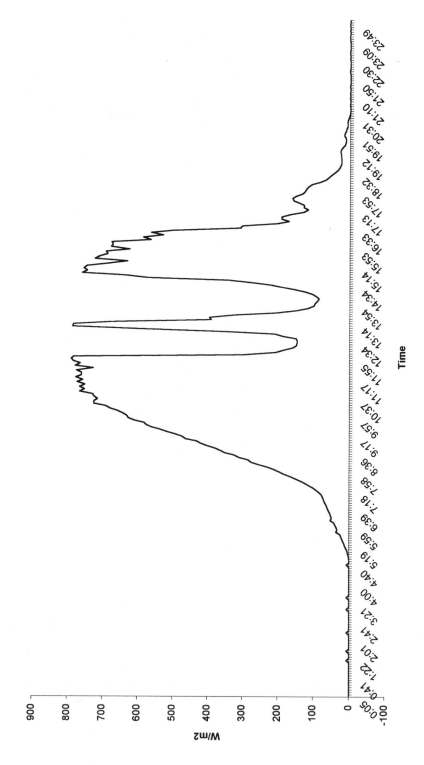

Figure 2.13   Global irradiance on a photovoltaic array, 21 July 1997
Data supplied by Dr T. Markvart, Southampton University

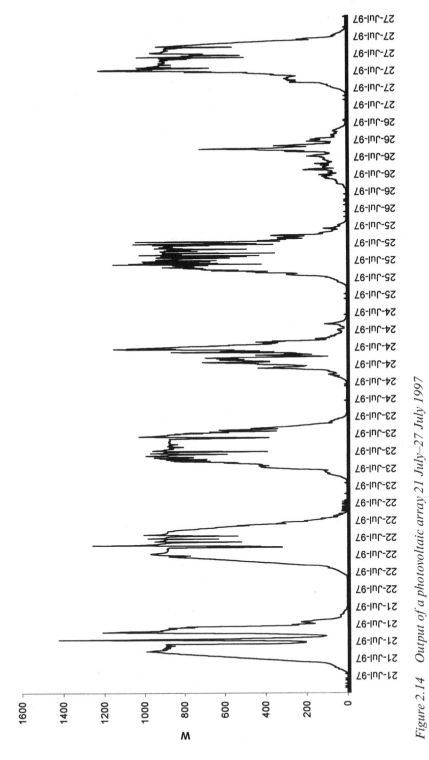

Figure 2.14 *Output of a photovoltaic array 21 July–27 July 1997*
Data supplied by Dr T. Markvart, Southampton University

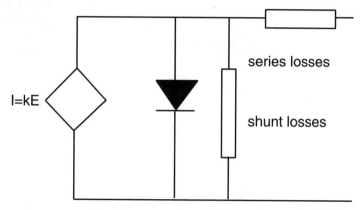

Current source output is proportional to irradiance ie I=kE.

Figure 2.15   Equivalent circuit of a PV cell

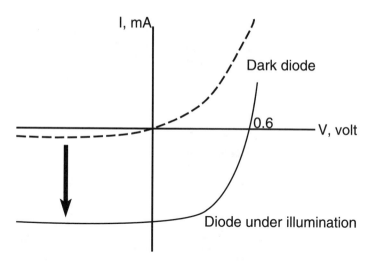

Figure 2.16   Characteristic of an illuminated diode

series/parallel strings and packaged into modules for mechanical protection. These robust and maintenance-free modules then have an electrical characteristic shown, by convention, as in Figure 2.17.

The maximum power output of the module is obtained near the knee of its characteristic. However, the output voltage, but not the current, of the module is reduced at increased cell temperature and so the maximum output power can only be obtained using a maximum power point tracking (MPPT) stage on the input to the converter.

Figure 2.18 shows the schematic diagram of an inverter for a small PV

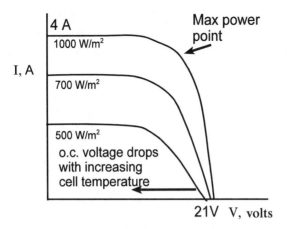

Figure 2.17   Typical characteristic of a PV module

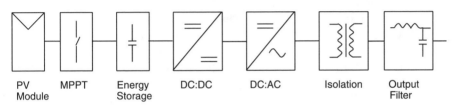

| PV Module | MPPT | Energy Storage | DC:DC | DC:AC | Isolation | Output Filter |

Figure 2.18   Schematic diagram of a small PV inverter for 'grid-connected' operation

'grid-connected' system. (Note that the term 'grid-connected' is often used rather loosely to describe small embedded generation systems which are connected to a local utility distribution network and not to the interconnected high-voltage interconnected grid network). The inverter typically consists of: (i) an MPPT circuit, (ii) an energy storage element, usually a capacitor, (iii) a DC:DC converter to increase the voltage, (iv) a DC:AC inverter stage, (v) an isolation transformer to ensure that DC is not injected into the network and (vi) an output filter to restrict the harmonic currents passed into the network, particularly those near the device switching frequencies. Very small inverters (up to say 200 W) may be fitted to the back of individual modules, the so-called 'AC module' concept, or rather larger inverters (up to say 3 kW) used for a number of modules, the 'string inverter' concept. It remains unclear which approach is likely to be the most cost-effective, and both types of inverter are available.

Although all photovoltaic cells operate on the same general principles there are a number of different materials used. The early cells used mono-crystalline silicon, and this is still in common use. Very large single crystals of silicon are formed and then cut into circular wafers and doped. The single crystal is expensive to form but allows high efficiencies, and overall module efficiencies up to 20% may be obtained. An alternative technique, again using bulk silicon, is to cast poly-crystalline cubes and then cut these into square wafers. Although this is a cheaper process, poly-crystalline modules are typically some 4% less efficient due to the random crystal structure of the cells. Both mono-crystalline and poly-crystalline silicon cells are in general use, and the choice between them is generally made on commercial grounds.

Bulk purified silicon is expensive, and therefore much effort has been expended on 'thin film' devices where only a small volume of active material is deposited on a cheaper, inert substrate. The active materials commonly used include amorphous silicon and cadmium telluride but a large number of other materials have been investigated. Thin film materials are less efficient than bulk silicon and, in the past, the performance of some cells degraded over time. However, thin film solar cells are extensively used in consumer products and are offered by a number of manufacturers for general application.

The widespread introduction of small photovoltaic systems mounted on, or even integrated into the fabric of, buildings will require different consideration compared to other larger forms of embedded generation (e.g. wind turbines, CHP units or hydro-sets). The photovoltaic units will have much smaller outputs (typically up to 1–2 kW) for a domestic dwelling and will be installed by home owners and their builders with minimal engineering input. It is clearly unreasonable to expect the owner of a small photovoltaic panel (rated at perhaps 250 W) to be subject to the same regulatory and administrative framework as the operator of a 10 MW CHP plant. This has been recognised and standards have been developed [17,18] specifically covering the connection of very small photovoltaic generators. which allow a much simpler approach than that required for larger embedded generation plant [19, 20].

2.3 Summary

The number of installations and different types of generating plant connected to the distribution network is constantly increasing. This embedded generation can offer considerable social and economic benefits, including high overall efficiencies and the exploitation of renewable energy sources, which lead to reduced gaseous emissions. The location and rating of the generators generally is fixed either by the heat load or by the renewable energy resource. Renewable energy generation is operated in response to the available resource to obtain the maximum return on the, usually high, capital investment. CHP plant is almost always operated in response to the needs of the host facility or a district heating load. Operation with regard to the needs of the power system is likely to reduce its efficiency unless, as in the case of the Danish rural CHP schemes, some form of energy storage is used. It is, of course, possible to install embedded generation, e.g. small gas turbines or reciprocating internal combustion engines, specifically to support the power system but this, so far, is uncommon.

When considering the connection of embedded generation it is essential to have a good understanding of the performance characteristics of the plant to ensure that the performance of the distribution system is not degraded. This is not an easy task given the range of types of plant and the different operating regimes adopted by their owners. If the embedded generation moves beyond being a simple source of kWh to the distribution network and starts to contribute to the operation of the power system then an even more detailed understanding of the prime movers and generators is required.

2.4 References

1 UK GOVERNMENT STATISTICAL OFFICE, 'Digest of UK Energy Statistics 1999' (The Stationery Office), ISBN 011 5154639
2 HORLOCK, J. H.: 'Cogeneration: combined heat and power thermo-dynamics and economics' (Pergamon Press, Oxford, 1987)
3 JORGENSEN, P., GRUELUND SORENSEN, A., FALCK CHRISTENSEN, J., and HERAGER, P.: 'Dispersed CHP units in the Danish Power System'. Paper no. 300–11, CIGRE Symposium on *Impact of demand side management, integrated resource planning and distributed generation*, Neptun, Romania, 17–19 September 1997
4 MARECKI, J.: 'Combined heat and power generating systems' (Peter Peregrinus, London, 1988)
5 KHARTCHENKO, N. V.: 'Advanced energy systems' (Taylor and Francis, Washington, 1998)
6 PACKER, J., and WOODWORTH, M.: 'Advanced package CHP unit for small-scale generation', *Power Engineering Journal*, 1991, **5**, (3), pp. 135–142
7 HU, S. D.: 'Cogeneration' (Reston Publishing Company, Reston, VA, 1985)
8 ALLAN C.L.C.: 'Water-turbine-driven induction generators'. Proceedings of the IEE, Part A, 3140S, December 1959, pp. 529–550
9 TONG JIANDONG, ZHENG NAIBO, WANG XIANHUAN, HAI JING, and DING HUISHEN: 'Mini-hydropower' (John Wiley and Sons, Chichester, 1997)
10 FRAENKEL, P., PAISH, O., BOKALDERS, V., HARVEY, A., BROWN, A., and EDWARDS, R.: 'Micro-hydro power, a guide for development workers' (IT Publications, London , 1991)
11 BOYLE, G. (Ed.): 'Renewable energy' (Oxford University Press, 1996)
12 FRERIS, L.L. (Ed.): 'Wind energy conversion systems' (Prentice Hall, 1990)
13 HEIER, S.: 'Grid integration of wind energy conversion systems' (John Wiley and Sons, Chichester, 1998)

14   GREEN, M.A.: 'Solar cells' (Prentice Hall, 1982)
15   VAN OVERSTRAETEN, R.J., and MERTENS, R.P.: 'Physics, technology and use of photovoltaics' (Adam Hilger, 1986)
16   MARKVART, T. (Ed.): 'Solar electricity' (Wiley, 1994)
17   ELECTRICITY ASSOCIATION: 'UK technical guidelines for inverter connected single phase photovoltaic (PV) generators up to 5 kVA'. Engineering Recommendation G77 Draft, 1999
18   NOVEM and EnergieNed: 'Guidelines for the electrical installation of grid connected photovoltaic (PV) systems' (EnergieNed and Novem, Holland, December 1998)
19   ELECTRICITY ASSOCIATION: 'Recommendation for the connection of embedded generation plant to Public Electricity Suppliers' distribution systems'. Engineering Recommendation G59/1, Amendment 2, 1994
20   ELECTRICITY ASSOCIATION: 'Recommendation for the connection of embedded generation plants to Public Electricity Suppliers' systems above 20 kV, or with outputs over 5 MW', Engineering Recommendation G75, 1994

Carno Wind Farm, Powys, mid-Wales
The wind farm consists of 56 600 kW, two speed, stall regulated wind turbines.
It is connected into the local utility at 132 kV.

(Source: National Wind Power)

Chapter 3

System studies

3.1 Introduction

An embedded generator is connected, by definition, to an electrical distribution network. This network is the conduit through which it exports the electrical energy that it produces. Since these exports can have a significant effect on the pattern of flows in the network, it is important to check that they will not degrade the quality of supply for the other users of the network. In most cases, this network was not designed for the sole use of the generator. It may indeed have been delivering power to consumers for many years before the embedded generator was commissioned. If the rating of this generator is a significant fraction of the capacity of the network, it will have a marked effect on the performance of this network. Reciprocally, this network can severely restrict the generator's ability to export power. An embedded generator must therefore be analysed as a component of a system. Its proponents and the owners of the distribution network must perform system studies to ascertain whether the network will need to be reinforced to accommodate the embedded generator. In some cases, these system studies may show that, rather than reinforcing the network, it may be more cost-effective to place limits or restrictions on the operation of the generator.

This chapter will first review the purposes of power flow computations, fault level calculations, stability studies and electromagnetic transient analysis. Then, the principles behind each of these types of studies will be explained. Examples of applications to realistic embedded generation installation will be presented and the data requirements will be discussed. This chapter should provide the user with the knowledge and understanding that are required to use confidently the software packages designed to perform these system studies.

3.2 Types of system studies

The design of distribution networks is driven by two fundamental goals: delivering an acceptable quality of supply to consumers under normal conditions and protecting the integrity of the system when the network is affected by faults.

Items of electrical equipment installed by consumers (and particularly electronic equipment) are intended to be operated within a relatively narrow voltage range around the nominal voltage. The voltage at all nodes of a distribution network must therefore remain within this range for all expected loading conditions. A power flow program (sometimes also called a load flow program) is the tool used for checking the normal operating states of an electrical power network.

A number of factors can damage a distribution network: strong winds or accumulation of ice can break overhead conductors, careless street digging can rupture cables and natural decay or rodents can weaken insulation. Such damage creates a fault or short circuit, i.e. an easier path for the current. Faults are not only a safety hazard but large fault currents can seriously damage equipment. Fault calculation programs are used to calculate the fault currents that would occur for different network configurations and fault locations. Their results are used not only to check that the components of the network have a sufficient rating to withstand the fault current but also to verify that these fault currents are sufficiently large for protection devices to detect the fault!

Power flows and sustained fault currents are calculated assuming that the system has reached a steady-state equilibrium, either normal or faulted. Faults, however, can perturb this equilibrium to such an extent that this steady-state assumption is no longer valid. The power system must then be treated as a dynamic system. Transient stability programs model the system using differential equations instead of algebraic equations. They are used to check whether all rotating machines continue to operate in synchronism following a disturbance. If synchronism is maintained, the system is deemed stable, otherwise it is considered unstable.

When fast transients must be studied or the behaviour of power electronic or other non-linear devices must be analysed, the three types of programs described above may be inadequate. These programs are indeed based on the assumption that the voltage and current waveforms are sinusoidal and that the system can be modelled using phasors. When this assumption is no longer valid or acceptable, the system must be modelled on a much smaller time-scale using an electromagnetic transient program. In this type of program, the waveforms are not assumed to be sinusoidal but are recomputed step-by-step using a detailed differential equation representation of all the components of the system. While such programs represent the behaviour of the system very accurately and in great detail, their use requires a significant amount of skill.

3.3 Power flow studies

Given all generations and loads in a system, a power flow calculation provides the voltage at all the nodes in this system. Once these voltages are known, calculating the flows in all the branches is straightforward. Power flow studies are simply the application of power flow calculations to a variety of load and generation conditions and network configurations. This section first explains the principles underlying power flow calculations using a simple two-bus example. These principles are then generalised to networks that are more complex and illustrated using examples from an embedded generation scheme.

3.3.1 Power flow in a two-bus system

Consider the one-line diagram of a two-bus power system shown in Figure 3.1. This system will be modelled as shown in Figure 3.2. The line's resistance and reactance are taken into account but its capacitance is neglected.

Suppose that the generator voltage $\overline{V}_G$ and the complex power $\overline{S}$ injected by this generator are known. We would like to calculate the voltage $\overline{V}_L$ at the load.

The complex power is related to the current and voltage by the following expression:

$$\overline{S}_G = P_G + jQ_G = \overline{V}_G \overline{I}^* \tag{3.1}$$

Figure 3.1   Two-bus power system

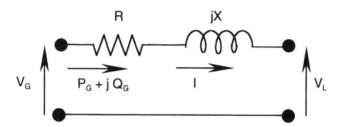

Figure 3.2   Model of a two-bus power system. The active and reactive powers are specified at the generator bus

From this expression we can find an expression for the current in terms of the injected power and the voltage:

$$\bar{I} = \frac{P_G - jQ_G}{V_G^*} \tag{3.2}$$

Using Kirchhoff's voltage law, the voltage at the load is given by

$$\overline{V_L} = \overline{V_G} - (R + jX)\bar{I} \tag{3.3}$$

Combining eqns. (3.2) and (3.3), we obtain

$$\overline{V_L} = \overline{V_G} - (R + jX)\frac{(P_G - jQ_G)}{V_G^*} \tag{3.4}$$

Assuming that the generator voltage is chosen as the reference for the phase angles,

$$\overline{V_G} = \overline{V_G^*} = V_G \angle 0° = V_G \tag{3.5}$$

eqn. (3.4) becomes

$$\overline{V_L} = V_G - \frac{RP_G + XQ_G}{V_G} - j\frac{XP_G - RQ_G}{V_G} \tag{3.6}$$

If we were given the values of the load and the desired voltage at the load bus and were asked to calculate the generator voltage, a similar procedure would yield the following relation:

$$\overline{V_G} = V_L + \frac{RP_L + XQ_L}{V_L} + j\frac{XP_L - RQ_L}{V_L} \tag{3.7}$$

where the load voltage has been taken as the reference for the angles.

Eqns. (3.6) and (3.7) show that if we are given the voltage and power injection at the same end of the line, an explicit formula will give us the voltage at the other end of this line. In practice, the voltage may be specified at the generator bus while the power is specified at the load as shown in Figure 3.3. The power consumed by the load differs from the power injected by the generator because the resistance and reactance of the line cause active and reactive power losses.

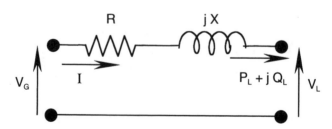

Figure 3.3   *Model of a two-bus power system. The active and reactive powers are specified at the load bus*

In this case, the complex power at the load bus is related to the voltage and current by

$$\overline{S_L} = P_L + jQ_L = \overline{V_L}\,\overline{I}^* \tag{3.8}$$

Extracting the current from the above expression and replacing in eqn. (3.3), we get

$$\overline{V_L} = \overline{V_G} - (R + jX)\frac{(P_L - jQ_L)}{V_L^*} \tag{3.9}$$

We cannot calculate the voltage at the load bus using eqn. (3.9) because this voltage appears in a non-linear fashion on both sides of the equation. The linear relations, which are the norm in circuit analysis, have disappeared because we have chosen to describe the load not in terms of impedances or currents but in terms of real and reactive power. This type of equation is usually solved using the following iterative algorithm:

Step 1 Guess an initial value $\overline{V_L^0}$ for the load bus voltage.

(If no information is available, assume that $\overline{V_L^0} = 1.0 \angle 0°$ p.u.)

Step 2 Compute $\overline{V_L^{i+1}} = \overline{V_G} - (R + jX)\dfrac{(P_L - jQ_L)}{V_L^{i*}}$

Step 3 If $|\overline{V_L^{i+1}} - \overline{V_L^i}| > \varepsilon$, where ε is a predefined tolerance, the iterative procedure has not yet converged. Let $i = i + 1$ and go back to Step 2.

Step 4 If this convergence condition is satisfied, the computation can be stopped.

Once the voltage at the load bus has been obtained using this iterative procedure, the current in the line can be calculated using eqn. (3.8). The active and reactive powers supplied by the generator are then equal to the active and reactive loads plus the active and reactive losses in the line:

$$P_G = P_L + I^2 R \tag{3.10}$$

$$Q_G = Q_L + I^2 X \tag{3.11}$$

Example 3.1: Suppose that, given $V_G = 1.0$ p.u., $P_L = 0.5$ p.u., $Q_L = 0.3$ p.u., $R = 0.01$ p.u. and $X = 0.1$ p.u., we wish to compute V_L with an accuracy of 0.001 using the algorithm described above. Table 3.1 shows that convergence is achieved after three iterations.

Table 3.1 Convergence record of the power flow calculation of Example 3.1

Iteration	$\overline{V_L^i}$	$\overline{V_L^{i+1}}$	$|\overline{V_L^{i+1}} - \overline{V_L^i}|$
0	$1.0 \angle 0°$	$0.966144 \angle -2.78837°$	0.058600
1	$0.966144 \angle -2.78837°$	$0.962590 \angle -2.78837°$	0.003554
2	$0.962590 \angle -2.78837°$	$0.962456 \angle -2.79900°$	0.000224

The current in the line is $I = 0.60584\angle-33.763°$ p.u. The powers injected by the generator are $P_G = 0.50367$ p.u. and $Q_G = 0.3367$ p.u. Note that since the reactive losses are larger than the active losses, the power factor at the generator is slightly smaller than the power factor of the load.

Note that power system computations are usually carried out in per unit (p.u.) rather than SI units. Transforming all quantities to the per unit system is a form of normalisation such that the nominal voltage of all voltage levels is equal to 1.0 p.u. Furthermore, after this transformation, all transformers are represented by impedances. A detailed discussion of the per unit system can be found in most introductory power system analysis textbooks such as References 1–3.

3.3.2 Relation between flows and voltages

Eqn. (3.6) can be written in the following form:

$$\overline{V}_L = V_G - \Delta V - j\delta V \qquad (3.12)$$

The component of the voltage drop in phase with the generator voltage is given by

$$\Delta V = \frac{RP_G + XQ_G}{V_G} \qquad (3.13)$$

and the component of the voltage drop in quadrature with the generator voltage by

$$\delta V = \frac{XP_G + RQ_G}{V_G} \qquad (3.14)$$

Figure 3.4 illustrates these relations. Note that this illustration is purely qualitative because the large voltage drop depicted in this figure would not be acceptable in a real power system. From this figure, we can observe that most of the difference between the magnitudes of the voltages at the generator and the load is due to ΔV, the in-phase component of the voltage drop. On the other hand, the entire phase angle difference is due to the quadrature component δV.

Since the resistance of transmission lines is usually negligible compared to their reactance, eqns. (3.9) and (3.10) can be approximated:

$$\Delta V \approx \frac{XQ_G}{V_G} \qquad (3.15)$$

$$\delta V \approx \frac{XP_G}{V_G} \qquad (3.16)$$

Changes in voltage magnitude across a line or a portion of a power system are therefore caused mostly by transmission of reactive power, while phase shifts between voltages are due mostly to the transfer of

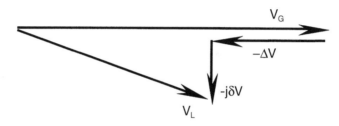

Figure 3.4   Components of the voltage drop across a transmission line

active power. Conversely, if there is a large difference in angle between the voltages at two buses connected by a line, there will be a large transfer of active power between these two buses. Similarly, a large difference in voltage magnitude between two buses results in a large transfer of reactive power. It should be noted that the resistance of distribution lines is usually not much smaller than their reactance. The approximations represented by eqns. (3.15) and (3.16) therefore may be inappropriate in distribution systems.

If a load is capacitive, the reactive power that it produces must be absorbed by the generator. In that case the direction of flow of the reactive power is reversed, and eqn. (3.15) suggests that the magnitude of the voltage at the load may actually be greater than the magnitude of the voltage at the generator.

3.3.3 Power flow in larger systems

In the case of the two-bus system, combining Kirchhoff's voltage law with the known power injection produced a formula for calculating the missing voltage. This ad hoc approach does not work if the system is more complex, i.e. if it contains more than one generator or load or if the network is meshed. A more systematic method is therefore needed.

This systematic approach relies on the observation that at each bus in the system the power must be in balance. In other words, the sum of the powers produced and consumed at the bus and of the powers transmitted to or from other buses by lines, cables or transformers must be zero. Figure 3.5 illustrates this idea and the concept of net active and reactive injections P_k and Q_k. Since the active and reactive powers are balanced separately, we have:

$$P_k = P_{Gk} - P_{Lk} = \sum_{i \in N_k} P_{ki}$$

$$Q_k = Q_{Gk} - Q_{Lk} = \sum_{i \in N_k} Q_{ki} \tag{3.17}$$

where the summation extends to all buses j that are neighbours of bus k (i.e. that are directly connected to bus k by a line, cable or transformer.)

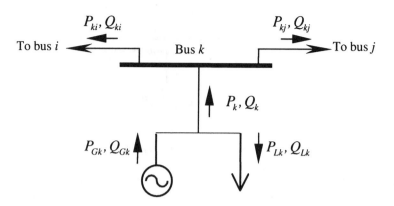

Figure 3.5   Balance of power at bus k

The next step in the systematic approach is to observe that the active and reactive flows in the network obey Kirchhoff's voltage law. If we replace the P_{ki} and Q_{ki} terms in eqns. (3.17) and (3.19) by expressions involving the voltages, we will have a set of equations relating voltages, power injected and power consumed at all buses. Using the notations shown in Figure 3.6, we have:

$$\begin{aligned}
\overline{I}_{ki} &= \frac{\overline{V}_k - \overline{V}_i}{R_{ki} + jX_{ki}} = \frac{\overline{V}_k - \overline{V}_i}{\overline{Z}_{ki}} = \overline{Y}_{ki}(\overline{V}_k - \overline{V}_i) \\
&= Y_{ki} \angle \delta_{ki}(V_k \angle \theta_k - V_i \angle \theta_i) \\
&= Y_{ki}V_k \angle (\theta_k + \delta_{ki}) - Y_{ki}V_i \angle (\theta_i + \delta_{ki})
\end{aligned} \tag{3.18}$$

Hence

$$\overline{I}_{ki}^* = Y_{ki}V_k \angle (-\theta_k - \delta_{ki}) - Y_{ki}V_i \angle (-\theta_i - \delta_{ki}) \tag{3.19}$$

and

$$\begin{aligned}
\overline{V}_k\, \overline{I}_{ki}^* &= Y_{ki}V_k^2 \angle (\theta_k - \theta_k - \delta_{ki}) - Y_{ki}V_kV_i \angle (\theta_k - \theta_i - \delta_{ki}) \\
&= Y_{ki}V_k^2 \angle (-\delta_{ki}) - Y_{ki}V_kV_i \angle (\theta_k - \theta_i - \delta_{ki})
\end{aligned} \tag{3.20}$$

Since:

$$\begin{cases} P_{ki} = \mathrm{Re}(\overline{S}_{ki}) = \mathrm{Re}(\overline{V}_k\, \overline{I}_{ki}^*) \\ Q_{ki} = \mathrm{Im}(\overline{S}_{ki}) = \mathrm{Im}(\overline{V}_k\, \overline{I}_{ki}^*) \end{cases} \tag{3.21}$$

we have:

$$\begin{cases} P_{ki} = Y_{ki}V_k^2\cos(-\delta_{ki}) - Y_{ki}V_kV_i\cos(\theta_k - \theta_i - \delta_{ki}) \\ Q_{ki} = Y_{ki}V_k^2\sin(-\delta_{ki}) - Y_{ki}V_kV_i\sin(\theta_k - \theta_i - \delta_{ki}) \end{cases} \tag{3.22}$$

Note that due to the losses caused by the resistance and reactance of the branch, the powers measured at the two ends of the branch are not equal:

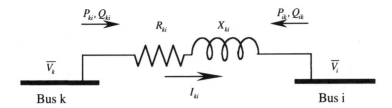

Figure 3.6   Notation used in the calculation of the flow in branch ki

$$P_{ki} \neq P_{ik}$$
$$Q_{ki} \neq Q_{ik}$$
(3.23)

Combining eqns. (3.17) and (3.22), we get:

$$\begin{cases} P_k = \sum_{i \in N_k} \{ Y_{ki} V_k^2 \cos(-\delta_{ki}) - Y_{ki} V_k V_i \cos(\theta_k - \theta_i - \delta_{ki}) \} \\ Q_k = \sum_{i \in N_k} \{ Y_{ki} V_k^2 \sin(-\delta_{ki}) - Y_{ki} V_k V_i \sin(\theta_k - \theta_i - \delta_{ki}) \} \end{cases}$$
(3.24)

These last equations relate the active and reactive power injection at a bus to the voltage magnitude and angle at this bus and at its neighbours. Since similar equations can be written for each of the *n* buses in the system, we have *2n* equations. These equations relate *4n* variables:

- *n* active power injections P_k
- *n* reactive power injections Q_k
- *n* voltage magnitudes V_k
- *n* voltage angles θ_k.

Two of these variables must therefore be given at each node to ensure a balance between the number of equations and the number of unknowns. Three combinations of known and unknown variables are used in practice. These combinations are related to the physical characteristics of the buses, as follows.

PQ buses are typically load buses where the net active and reactive power injections are known, while the voltage magnitudes and angles are unknown.

PV buses are generator buses where the automatic voltage regulator keeps the voltage magnitude at a constant level by adjusting the field current of the generator and hence its reactive power output. The active power output of the generator and the voltage magnitude are therefore known, while the reactive injection and the voltage angle depend on the loading of the system.

All power flow calculations should include one (and only one) *reference bus or slack bus.* This bus has a dual purpose. First, as in all AC circuit calculations, the phase angle of one quantity can be chosen arbitrarily. The angle of all other voltage and current phasors is calculated relative to this reference. In power flow calculations, the voltage angle at one bus (the reference bus) is therefore arbitrarily set to zero. Secondly, provision needs to be made for the losses in the system. The sum of the active powers produced by the generators must be equal to the sum of the active powers consumed by the loads plus the active power losses in the transmission network. These losses depend on the currents in the branches of this network, which in turn depend on the voltages, and calculating these voltages is the object of the power flow calculation! To resolve this problem, the active power injection at one bus (the slack bus) is therefore left unspecified. The reference bus is usually chosen as the slack bus. During the power flow calculation, the angle at this bus is therefore kept constant while the active power injection is allowed to vary to compensate for any change in the losses. In a transmission network, a bus with a large generating unit is usually chosen as the slack/reference bus. In a distribution network, the bus where the distribution system under investigation is connected to a higher voltage level is usually chosen as the slack/reference bus. Since the voltage magnitude at such buses is normally specified, the reactive injection is unknown. The slack/reference bus can therefore be described as a θV bus.

Note that the selection of known and unknown variables at the various types of buses follows the active power/voltage angle and the reactive power/voltage magnitude pairings described above.

A variety of control devices can be installed to improve the operating characteristics of a power system. On-load tap changing transformers, voltage regulators and a static VAr compensator (SVC) attempt to maintain the voltages within the acceptable range while excitation limiters protect the generators. These devices have a significant effect on the steady-state behaviour of power systems. They must therefore be considered during power flow studies. A description of the techniques used for modelling these devices is beyond the scope of this chapter but can be found in Reference 4.

3.3.4 Solving the power flow equations

The power flow equations, eqns. (3.24), are non-linear and cannot be solved manually except for trivial systems. Sophisticated iterative methods have been developed for solving them quickly and accurately. Detailed descriptions of these methods can be found in References 1–3. Many software packages specially designed for carrying out power flow studies are available on the market.

Running power flow studies with the help of a commercial package involve the following steps.

Gathering data: This is often the most time-consuming task. The impedances of the lines and cables must be calculated based on the data provided by the manufacturer and the layout of the network. The parameters of the generators and transformers must be extracted from the relevant data sheets. Data pertaining to the distribution network to which the embedded generation plant will be connected must be obtained from the operator of this network. All quantities must then be converted to a consistent per unit system.

Creating a model: The data gathered at the previous step are then used to create a model of the system to be studied. Older power flow programs require the user to enter these data in a file according to a precisely defined format. With modern programs, the user draws a diagram of the network before entering the parameters through forms.

Setting up cases: The user must then decide the load and generation conditions for which a power flow must be calculated. These data, as well as the position of the transformer taps and the settings of other control devices, must also be provided for the program.

Running the program: This is the easy part, unless the iterative method does not converge! Divergence is usually caused by errors in the model. There is unfortunately no easy way to determine which parameters are faulty. Errors in the network topology are easily made when this information is not entered through a graphical user interface. The value of all the components of the model must be checked. Excessive loading can also be a cause of divergence. This form of non-convergence is an indication that voltages in the system would be unacceptably low for loads in that range.

Analysing the results: Once a solution has been obtained, it must be checked for reasonableness. Slightly surprising results should be investigated carefully as they can be an indication of a minor error in the model. Once the user is satisfied that the model is correct, the program can be used to study other load and generation conditions as well as other network configurations.

3.3.5 Application to an embedded generation scheme

Figure 3.7 represents the essential features of a distribution network into which a generator is embedded at bus D. The connection of this network to the transmission grid is represented by a single generator in series with a transformer supplying bus A. Since bus S has been chosen as the slack bus, this generator will supply the active power required to balance the system. Table 3.2 gives the parameters of the lines and transformers. The voltage at bus S is assumed to be held constant at its nominal value by the source generator, while the tap-changers on the transformers between buses A and B maintain the voltage at bus B at its nominal value.

Let us first consider the case where the system is supplying its maximum load when the embedded generator is not producing any power. Figure 3.8 summarises the voltages, injections and flows that have been calculated for these conditions using a commercial power flow program. These results demonstrate that this system is relatively 'weak': the voltage

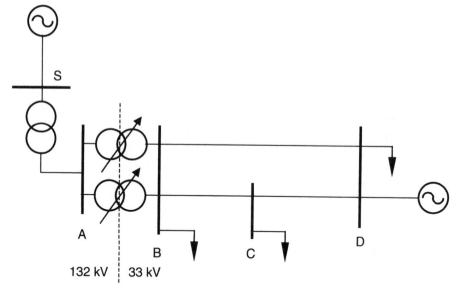

Figure 3.7   Portion of a distribution system with an embedded generator. This network is connected to the transmission system at bus S and the embedded generator is connected at bus D

Table 3.2   Parameters for the network used for power flow studies

From bus	To bus	Type	Resistance	Reactance
S	A	Transformer	0.0	0.06670
A	B	Transformer	0.00994	0.20880
A	B	Transformer	0.00921	0.21700
B	C	Line	0.04460	0.19170
B	D	Line	0.21460	0.34290
C	D	Line	0.23900	0.41630

All values are per unit

at bus D (0.953 p.u.) is marginally acceptable even though the voltage at bus B is held at its nominal value through the action of the tap-changing transformers. The active and reactive losses are also quite significant. It should be noted that these losses cause a difference between the active and reactive flows at the two ends of the lines and transformers. To keep the figure readable, only one value is given for these flows. This explains why the power balance may not appear to be respected at all buses in these figures.

If the embedded generator produces 20 MW at unity power factor, Figure 3.9 shows that the voltage profile is much more satisfactory. The

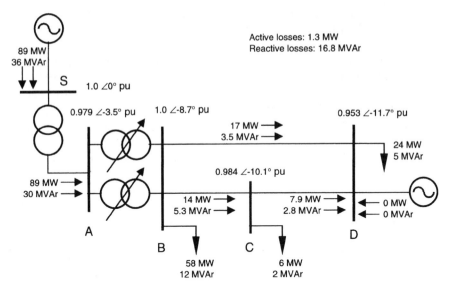

Figure 3.8   Power flow for maximum load and no embedded generation

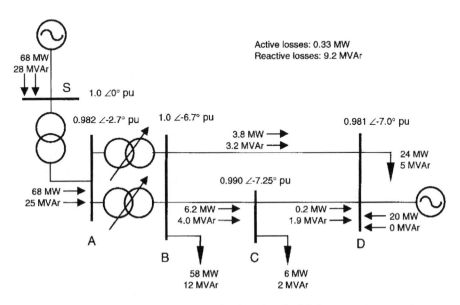

Figure 3.9   Power flow for maximum load and embedded generation at unity power factor

losses are considerably reduced because the generation is much closer to the load and the lines carry much reduced flows.

A further reduction in losses and an even better voltage profile can be achieved if, instead of operating at unity power factor, the embedded generator produces some reactive power. This case is illustrated in Figure 3.10. It is interesting to note that, under these circumstances, the active and reactive powers flow in opposite directions on two of the lines.

On the other hand, if, as shown in Figure 3.11, the embedded generator consumes reactive power (as an induction generator always does), the voltage profile and the losses are somewhat worse than in the unity power factor case.

If the embedded generator continues to produce its nominal power during periods of minimum load, the local generation may exceed the local consumption. In such cases, the pattern of flows is reversed and the distribution network injects power into the transmission grid. This case is illustrated by Figure 3.12, where the loads have been set at 10% of the maximum. The voltage phasor at bus D is not only the largest in magnitude but also leads all the other voltages. The source generator (which represents the rest of the system) absorbs the excess generation but supplies the necessary reactive power.

Some distribution network operators have expressed concerns that a reversal in the normal direction of flows caused by the presence of an embedded generator could interfere with the voltage regulation function

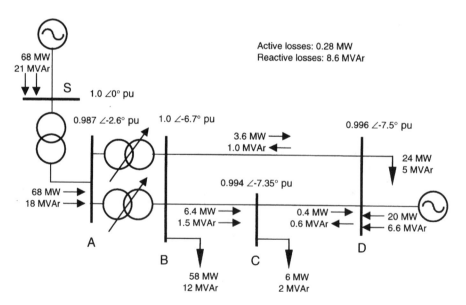

Figure 3.10　Power flow for maximum load and embedded generation at 0.95 power factor lagging (producing reactive power)

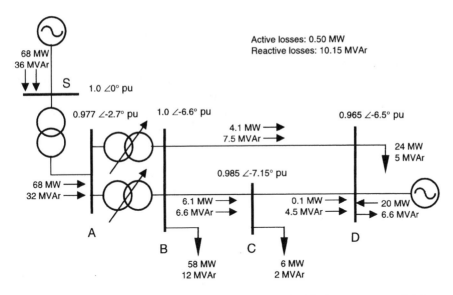

Figure 3.11   Power flow for maximum load and embedded generation at 0.95 power factor leading (absorbing reactive power)

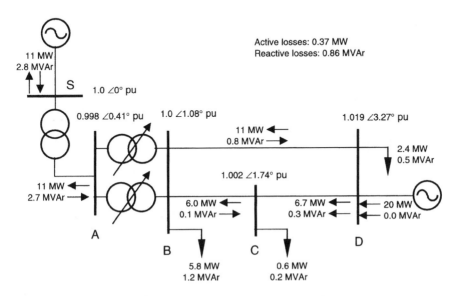

Figure 3.12   Power flow for minimum load and embedded generation at unity power factor

of tap-changing transformers. To investigate these concerns, the embedded generator has been moved to the secondary side of a 33/11kV, tap-changing transformer. The automatic voltage controller of this transformer has been set to control the 11 kV busbar voltage only. Any current compounding scheme (e.g. line drop compensation or negative reactance compounding) may be adversely affected by embedded generators. Figure 3.13 illustrates the case where the embedded generator produces not only 20 MW but also 6.6 MVAr for the minimum load conditions. It can be seen in this figure that the active and reactive powers flow from the embedded generator to the transmission grid. Even under these unusual conditions, the voltage at the 11 kV bus E is kept at its nominal value by the tap-changing transformer.

3.4 Fault studies

Conductors in distribution networks are separated from earth and from each other by a variety of insulating materials: air, paper or polymers. Occasionally, an unpredictable event ruptures this insulation, creating a short circuit between conductors or between conductors and

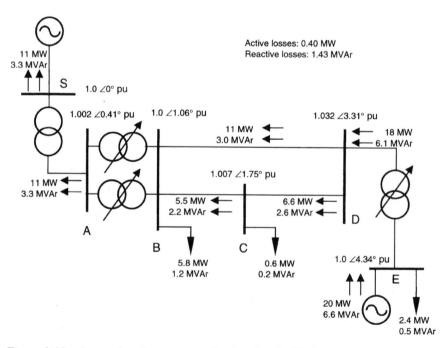

Figure 3.13   Power flow for minimum load and embedded generation at 0.95 power factor lagging on a voltage controlled bus

earth. This abnormal conducting path is called a fault. Being able to predict the value of the current in faults is very important for two main reasons. First, this current may be so large that it could damage the distribution plant or exceed the rating of the breakers that are supposed to interrupt it. Paradoxically, the second reason for calculating fault currents is to check that they are not too small for the fault to be detected. Devising a protection system capable of discriminating between a large (but normal) load current and a small fault current is difficult. Since failing to detect a fault is an unacceptable safety risk, the distribution system must be designed in such a way that fault currents are large enough to be detected under all operating conditions.

A distinction must be made between balanced and unbalanced faults. Balanced faults affect all three phases of the network in a similar manner and the symmetry between the voltages and currents in the three phases is not altered. A single-phase representation of the network can therefore be used when studying such faults. On the other hand, unbalanced faults create an asymmetry in the network and require a more complex analysis based on symmetrical components.

This section begins with an explanation of balanced fault calculations using a simple two-bus example. These calculations are then generalised to networks of arbitrary size and complexity and then to unbalanced faults. Finally, these concepts are illustrated using examples from an embedded generation scheme.

3.4.1 Balanced fault calculations

Let us consider again a simple two-bus power system such as the one shown in Figure 3.14. We will assume that no load is connected to this system. For balanced fault calculations, this system must be modelled as shown in Figure 3.15. The main difference between this model and the model used for power flow studies is that the generator is now represented as an ideal voltage source behind a source impedance Z_S. The total impedance of the line is represented by Z_L.

If a balanced fault occurs at the far end of the line, Figure 3.15 shows that the fault current is given by

$$\overline{I}_f = \frac{\overline{E}}{Z_S + Z_L} \tag{3.25}$$

On the other hand, if the fault occurs at a distance d down the line, the fault current is given by

$$\overline{I}_f = \frac{\overline{E}}{Z_S + \dfrac{l}{d} Z_L} \tag{3.26}$$

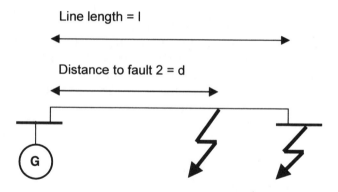

Figure 3.14   Two-bus power system for fault calculations

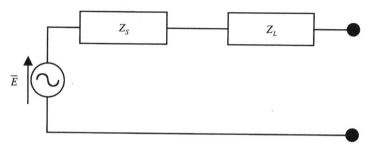

Figure 3.15   Model of the two-bus power system for fault calculations

From these equations, we conclude that the amount of impedance between the fault and the source determines the fault current. A short circuit at the end of a long distribution feeder will therefore result in a much smaller fault current than a fault near a primary substation.

In this very simple example, the expression for the fault current could be written by inspection. This is not the case in an actual system where the fault current might be supplied from multiple generators through a meshed network. In addition, it may be necessary to consider the effect of the prefault load current. As with the power flow, a solution to this problem arises from a more advanced technique of circuit analysis, namely Thevenin's theorem. This theorem states that an electrical circuit can always be modelled as an ideal voltage source in series with an impedance. For fault analysis, we need to model the network as seen from the location of a potential fault. The source voltage, which is called the Thevenin voltage and is denoted by $\overline{V}_{th}$, is equal to the voltage at that point before the occurrence of the fault. The series impedance, which is called the Thevenin impedance, is the impedance seen from the location of the fault looking back into the network. When we are interested *only* in what happens at a given location, a network of arbitrary complexity can therefore always be represented as shown in Figure 3.16. The fault current is then given by

$$\bar{I}_f = \frac{\overline{V}_{th}}{Z_{th}} \tag{3.27}$$

If the network under consideration is not too large or complex, the Thevenin impedance can be calculated by hand through network reduction. Furthermore, if a high degree of accuracy is not required, the Thevenin voltage can be assumed equal to the nominal voltage. On the other hand, for large networks where the prefault load current is not negligible, determining this voltage requires a power flow calculation. A systematic determination of the Thevenin impedance is best achieved through a partial inversion of the admittance matrix of the network. Specialised software packages have been developed to carry out such computations accurately and efficiently. Using such a package, determining the fault current for any number of potential fault locations, for several configurations of the network, and for various combinations of generating units is relatively easy.

3.4.2 Concept of fault level

Describing the effect of faults on a system in terms of the current that would flow in a fault at various locations could be somewhat confusing. This fault current must indeed be compared to the normal load current, and this load current is inversely proportional to the nominal voltage. To compensate for the effect of the voltage level, the magnitude of potential faults in the system is given in terms of the *fault level*. This quantity is usually expressed in MVA and is defined as

$$FL = \sqrt{3}\,V_{nominal}\,I_f \quad \text{(MVA)} \tag{3.28}$$

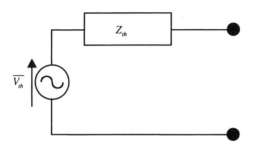

Figure 3.16   Thevenin equivalent of a network of arbitrary complexity

The base quantities for MVA, voltages and currents are usually chosen in such a way that they obey the following equation:

$$MVA_B = \sqrt{3}\, V_B I_B \tag{3.29}$$

If we divide eqn. (3.28) by eqn. (3.29) and assume that the base voltage is equal to the nominal voltage, we find that the per unit value of the fault level is equal to the per unit value of the fault current:

$$FL^{pu} = I^{pu} \tag{3.30}$$

Finally, if we assume that the voltage was equal to its nominal value prior to the fault, combining eqns. (3.27) and (3.30) gives

$$FL^{pu} = I^{pu} = \frac{1}{|Z_{th}^{pu}|} \tag{3.31}$$

The fault level therefore gives an indication of 'how close' a particular point is from the sources of a system. For example, fault levels in an EHV transmission system can be three orders of magnitude larger than in an LV distribution system. The configuration of the system can have a significant effect on the fault level. In particular, synchronising additional generators or connecting lines in parallel reduces the equivalent Thevenin impedance of the network and hence increases its fault level. Rather than providing a detailed model of their network to the designers of an embedded generation scheme, distribution utilities will usually give them the fault level at the connection point and the ratio X/R of the source impedance. Eqn. (3.31) shows that this information is sufficient to create a Thevenin equivalent of their network and hence to carry out simple fault studies.

Example 3.2: Figure 3.17 shows an embedded synchronous generator connected to a large power system through a distribution network. The fault level at the near end of this network is 1000 MVA at zero power factor (i.e. $X/R = \infty$). The impedance of the feeder is j0.1 p.u. while the source impedance of the synchronous generator is j0.2 p.u. All these impedances have been calculated using a 100MVA basis. If the network is operating at nominal voltage and the load current is negligible, what is

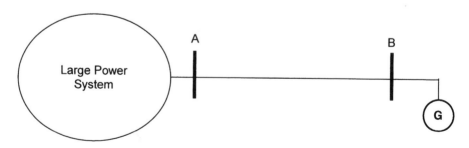

Figure 3.17   Power system of Example 3.2

the magnitude of the current that would flow if faults were to occur at the ends of the feeder?

Figure 3.18 shows the equivalent circuit that will be used for this calculation. The large power system has been replaced by its Thevenin equivalent with the impedance obtained from eqn. (3.31). Since the network operates at nominal voltage and the load current is assumed negligible, both voltage sources have been set at 1.0 p.u. The magnitude of the fault current at bus A is given by

$$I_{f,A} = \frac{1.0}{X_A} = \frac{1.0}{0.075} = 13.3 \text{ p.u.}$$

where X_A is the parallel combination of 0.1 p.u. and 0.3 p.u.

The magnitude of the fault current at B is

$$I_{f,B} = \frac{1.0}{X_B} = \frac{1.0}{0.1} = 10.0 \text{ p.u.}$$

where X_B is the parallel combination of 0.2 p.u. and 0.2 p.u.

In addition to its role in fault current calculations, the concept of fault level is also useful when analysing a network operating under normal conditions. As eqn. (3.31) indicates, larger fault levels correspond to smaller Thevenin impedances. The voltage drop caused by a given load current will therefore be smaller at buses where the fault level is high. Fault levels are thus sometimes used to quantify the 'strength' or 'stiffness' of a network at a given location.

3.4.3 Application to an embedded generation scheme

We will now use the small system introduced in Section 3.3.5 to illustrate the concepts developed in the previous sections. Let us first consider the case of Figure 3.19 where the embedded generator is disconnected from the system. The fault levels have been calculated (using a commercial-grade program) for faults at various buses in the system and are shown

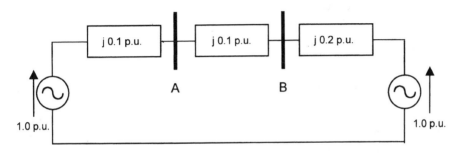

Figure 3.18   Model of the power system of Example 3.2

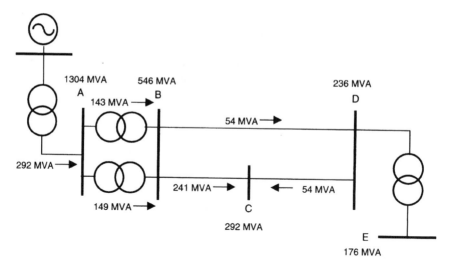

Figure 3.19   *MVA fault levels for balanced three-phase faults at the various buses in the system. The arrows show the MVA flows for a fault at bus C. The system is assumed to be unloaded prior to the fault*

next to the bus names. It is clear that the fault level decreases as the distance between the fault and the source increases. This figure also shows the flows that would result from a fault at busbar C. Note that the sum of the branch flows at bus C is only roughly equal to the fault level at this bus because the flows are expressed in MVA and the corresponding currents have slightly different phases. Figure 3.20 shows that the presence of an embedded generation significantly increases the fault levels in the system. In particular, 58 MVA would be drawn from this generator by a fault at bus C.

3.4.4 Unbalanced faults

Since balanced faults affect all three phases in an identical manner, the symmetry of the network is preserved and the fault current is balanced, albeit larger than a normal load current. The network impedances that must be considered in the calculation of balanced fault currents are therefore the 'normal' impedances of the network. A majority of the faults that occur in a network, however, do not affect all three phases in the same manner. Such faults are called unbalanced because they destroy the three-phase symmetry of the network. Unbalanced faults usually involve a short circuit between one line and the earth, between two lines or between two lines and the earth. Since the currents that result from such faults are not symmetrical, carrying out the analysis in terms of the

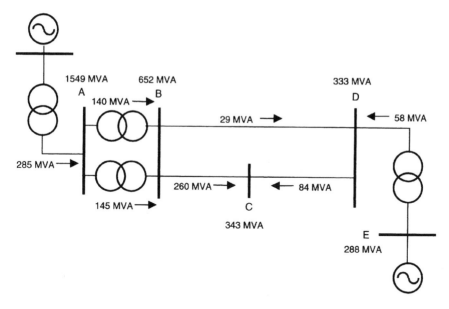

Figure 3.20   *MVA fault levels for balanced three-phase faults at the various buses in the system, including the contribution of the embedded generator at bus E. The arrows show the MVA flows for a fault at bus C. The system is assumed to be unloaded prior to the fault*

actual phase quantities is extremely difficult. Transforming the actual phase quantities into a set of abstract variables called symmetrical components considerably simplifies the calculation of unbalanced fault currents. It can be shown [5] that any set of unbalanced phase quantities (voltages or currents) can be decomposed into three components:

1   a positive sequence component consisting of three balanced voltages or currents in a normal (positive) phase sequence
2   a negative sequence component consisting of three balanced voltages or currents in a reverse (negative) phase sequence
3   a zero sequence component consisting of three voltages or currents of equal magnitude and phase.

The symmetrical component currents are related to the phase currents by the following matrix transformation:

$$\begin{bmatrix} \overline{I_0} \\ \overline{I_1} \\ \overline{I_2} \end{bmatrix} = \frac{1}{3} \begin{bmatrix} 1 & 1 & 1 \\ 1 & a & a^2 \\ 1 & a^2 & a \end{bmatrix} \cdot \begin{bmatrix} \overline{I_a} \\ \overline{I_b} \\ \overline{I_c} \end{bmatrix} \tag{3.32}$$

where a is the complex number,

$$a = 1\angle 120° = -\tfrac{1}{2} + j\frac{\sqrt{3}}{2} \tag{3.33}$$

The inverse transformation gives the phase currents in terms of the symmetrical component currents:

$$\begin{bmatrix} \overline{I_a} \\ \overline{I_b} \\ \overline{I_c} \end{bmatrix} = \begin{bmatrix} 1 & 1 & 1 \\ 1 & a^2 & a \\ 1 & a & a^2 \end{bmatrix} \cdot \begin{bmatrix} \overline{I_0} \\ \overline{I_1} \\ \overline{I_2} \end{bmatrix} \tag{3.34}$$

An identical transformation relates phase voltages and symmetrical component voltages.

The impedance that lines, cables, generators and transformers present to the flow of currents in a balanced system is the positive sequence impedance. The impedances that these devices offer to negative and zero sequence currents are called, respectively, the negative and zero sequence impedances.

In our discussion of balanced faults, we argued that, when observed from a given location, a network of arbitrary complexity could be replaced by its Thevenin equivalent. In that case, the Thevenin voltage source and impedance can be calculated from the positive sequence voltages produced by the generators and the positive sequence impedances of the components of the network. When unbalanced conditions are considered, we also need to take into account the equivalent circuits for the negative and zero sequence components. Since generators produce only a positive sequence voltage, there are no sources in the negative and zero sequence equivalent circuits. Figure 3.21 illustrates these concepts. A network of arbitrary complexity is examined from a given location (e.g. a potential fault location). Instead of considering the A, B and C phases of this network, we replace it by its positive, negative and zero sequence equivalent networks as seen from this location. Under normal conditions, no currents flow in these circuits and a voltage is present only in the positive sequence circuit. As discussed in Section 3.4.1, a balanced fault to the system at that location can be modelled by applying a short circuit on the equivalent circuit of the positive sequence. Each type of unbalanced fault can be modelled by a particular connection of the three equivalent circuits. For example, suppose that we would like to calculate the fault current that would result from a fault between phase A and earth. Since phases B and C are unaffected by this fault, we have:

$$\overline{I_b} = 0$$
$$\overline{I_c} = 0 \tag{3.35}$$

Inserting these values in eqn. (3.32), we obtain:

$$\begin{bmatrix} \overline{I_0} \\ \overline{I_1} \\ \overline{I_2} \end{bmatrix} = \tfrac{1}{3}\begin{bmatrix} 1 & 1 & 1 \\ 1 & a & a^2 \\ 1 & a^2 & a \end{bmatrix} \cdot \begin{bmatrix} \overline{I_a} \\ 0 \\ 0 \end{bmatrix} = \tfrac{1}{3}\begin{bmatrix} \overline{I_a} \\ \overline{I_a} \\ \overline{I_a} \end{bmatrix} \tag{3.36}$$

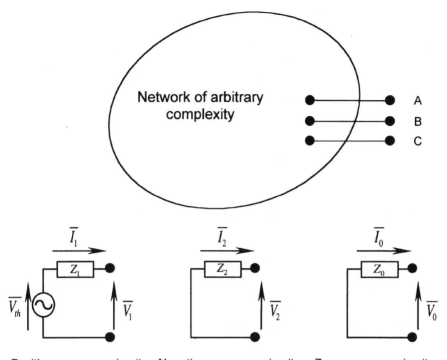

Positive sequence circuit   Negative sequence circuit   Zero sequence circuit

Figure 3.21   Equivalent circuits for unbalanced faults

from which we deduce that

$$\overline{I_0} = \overline{I_1} = \overline{I_2} = \tfrac{1}{3}\overline{I_a} \tag{3.37}$$

Figure 3.22*a* shows how eqn. (3.37) is translated into a connection between the symmetrical component circuits. From this figure it is then easy to see that the fault current is given by

$$\overline{I_a} = 3\overline{I_0} = 3\overline{I_1} = 3\overline{I_2} = \frac{3\overline{V_{th}}}{Z_1 + Z_2 + Z_0} \tag{3.38}$$

It can be shown [5] that line-to-line and line-to-line-to-earth faults can be modelled using the connections shown in Figures 3.22*b* and 3.22*c*. Using these diagrams and eqn. (3.34), we get the following relation for a fault between phases B and C:

$$\overline{I_a} = 0$$
$$\overline{I_b} = -\overline{I_c} = \frac{-j\sqrt{3}\,\overline{V_{th}}}{Z_1 + Z_2} \tag{3.39}$$

Similarly, for a fault involving phases B, C and earth:

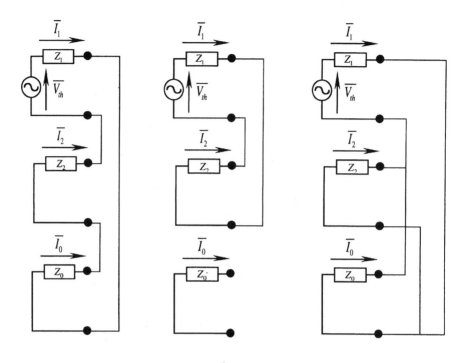

a) Line-to-earth fault   b) Line-to-line fault   c) Line-to-line-to-earth fault

Figure 3.22   *Connection of the symmetrical component equivalent circuits for line-to-earth, line-to-line and line-to-line-to-earth faults*

$$\overline{I}_a = 0$$

$$\overline{I}_b = \frac{\overline{V}_{th}}{Z_1 Z_2 + Z_1 Z_0 + Z_0 Z_2}[(1 - a)Z_2 + (1 - a^2)Z_0]$$

$$\overline{I}_c = \frac{\overline{V}_{th}}{Z_1 Z_2 + Z_1 Z_0 + Z_0 Z_2}[(a - a^2)Z_2 + (a - a^2)Z_0]$$

(3.40)

Effect of neutral earthing

The connection of generator and transformer windings and the earthing of neutral points play a particularly important role for line-to-earth faults. Figure 3.23 shows the three possible combinations of winding configurations and neutral earthing. For the delta and ungrounded Y connections, applying Kirchhoff's current law around the windings gives

$$\overline{I}_a + \overline{I}_b + \overline{I}_c = 0$$

(3.41)

while for the grounded Y connection we get

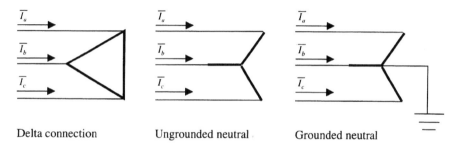

Delta connection   Ungrounded neutral   Grounded neutral

Figure 3.23   Types of transformer and generator winding connections

$$\overline{I}_a + \overline{I}_b + \overline{I}_c = \overline{I}_n \tag{3.42}$$

The first row of the matrix eqn. (3.32) gives

$$\overline{I}_0 = \tfrac{1}{3}(\overline{I}_a + \overline{I}_b + \overline{I}_c) \tag{3.43}$$

Comparing eqns. (3.41) and (3.43) we conclude that a zero sequence current cannot flow in windings that are in a delta or ungrounded Y connection. The zero sequence equivalent Thevenin impedance of generators and transformers whose windings are delta- or Y-connected is therefore infinite. On the other hand, eqn. (3.42) shows that a zero sequence current can flow in a grounded Y connection. Therefore, only the parts of the network that allow the circulation of a current through the earth should be included in the zero sequence equivalent circuit. If a fault occurs in a part of the network where there is no neutral-to-earth connection, the zero sequence equivalent impedance is infinite. Eqn. (3.38) and Figure 3.22a show that in such a case, a line-to-earth fault would not result in any fault current. Since such a fault would be difficult to detect, such a situation is extremely hazardous. In most countries, networks must therefore be designed in such a way that a neutral point is connected to earth under all conceivable configurations.

Unbalanced faults are more common but usually less severe than three-phase faults. However, when the neutral point of a generator is connected to earth through a zero or low-impedance connection, the current in a single line-to-earth fault can exceed the current in a balanced three-phase fault. This can also occur on the Y-grounded side of a Δ–Y grounded transformer.

3.4.5 Application to an embedded generation scheme

To compute the fault levels for unbalanced faults, we must know the configuration of the transformer windings. Figure 3.24 gives that information for the small system that we have already used in previous examples. This combination of winding configurations is typical of what

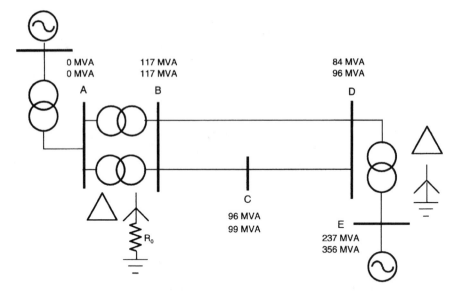

Figure 3.24   *MVA fault levels for single line-to-ground faults at various buses in the system. The upper number corresponds to the case where the embedded generator is not connected to the system and the lower number to the case where it is. The system is assumed to be unloaded prior to the fault*

we might find in a distribution system with embedded generation. The low-voltage side of the 132/33 kV transformers connecting buses A and B is star-connected and the neutral point is connected to earth through a resistance R_0. The value of R_0 is selected so that a line-to-ground fault on the low-voltage side would give rise to a 1000 A fault current. In addition, we must know, for each system component, not only its positive sequence impedance but also its negative and zero sequence impedance. While the negative sequence impedance of lines, cables and transformers is often assumed equal to the positive sequence impedance, the zero sequence impedance is usually significantly different. Table 3.3 gives these values for the relevant components of the small system of our example. Figure 3.24 gives two values of the fault level at each bus. The upper one corresponds to the case where the embedded generator at bus E is not connected to the system and the lower one to the case where it is connected. We observe that, contrary to the situation for balanced faults, the embedded generator does not significantly increase the fault level, except in the close vicinity of bus E. This is due to the fact that the fault current at buses B, C and D is limited mostly by the equivalent zero sequence impedance, which is dominated by the resistance R_0. A single line-to-ground fault at bus A would not produce any fault current as there is no path to earth represented in that part of the network. Finally,

Table 3.3   Zero sequence impedances for the network used for unbalanced fault studies

From bus	To bus	Type	R_0	X_0
A	B	Transformer	5.0250	0.2088
A	B	Transformer	5.0250	0.2170
B	C	Line	0.1440	0.9248
B	D	Line	0.3755	1.5920
C	D	Line	0.4397	2.0050

All values are per unit

it is interesting to note that the fault level at bus E is higher for a single line-to-ground fault than for a balanced three-phase fault.

3.4.6 Standards for fault calculations

The purpose of the previous sections was to introduce the basic concepts and methods used for calculating fault levels and fault currents. The reader should be aware that certain standards must be followed when these calculations are performed to design an embedded generation scheme. IEC 909 [6] shows how manual calculations may be performed, while Engineering Recommendation G74 [7] provides the basis for computer-based methods.

3.5 Stability studies

A power system is at equilibrium when the voltage magnitude and angle at each bus are such that power flows from buses where there is an excess of generation over demand to buses where demand exceeds supply. At each bus, there is thus a balance between the power generated, the power consumed and the power transmitted to and from other buses. Power flow programs calculate this equilibrium. This balance applies also to the generating units: the mechanical power provided by the prime mover is equal to electrical power produced by the generator, if we ignore the losses. In mechanical terms, this implies that the accelerating torque applied to the shaft by the prime mover is equal to the decelerating torque caused by the production of electrical power in the generator. Since the net torque is zero, the shaft rotates at constant speed. The angular position of the rotor is usually measured in a reference frame that rotates at synchronous speed. It is then called the rotor angle. At equilibrium, the rotor angle is a measure of the amount of power injected by the generator into the network.

In practice, a power system is never in the steady state as the loads and generations are constantly changing. In most cases, the voltages adjust themselves naturally to restore the equilibrium and the power system is said to be stable with respect to these small perturbations. On the other hand, a major disturbance such as a fault can induce an instability in the system and trigger a sequence of events leading to the collapse of part of the network.

The purpose of stability studies is therefore to verify that the system is designed in such a way that it can withstand the most severe credible disturbance. Conversely, stability studies are also used to determine the operating limits of an existing system. This section first develops a simple model of the mechanical behaviour of a synchronous generating unit. This generator is then connected in a quasi-radial manner to a very large power system. It is then shown how the equations of the mechanical and electrical subsystems can be used to study the transient stability of the system. Finally, an example shows how these principles can be applied to the study of an embedded generation scheme.

3.5.1 A simple dynamic model of the mechanical subsystem

The mechanical speed of a generator is described by Newton's equation for rotating masses,

$$J\frac{d^2\theta_m}{dt^2} = T_m - T_e \qquad (3.44)$$

where J is the combined moment of inertia of the generator and its prime mover

θ_m is the angular position of the shaft

T_m is the accelerating torque applied to the shaft by the prime mover

T_e is the reaction torque applied by the generator to the shaft.

The mechanical power supplied by the prime mover is transformed in electrical power by the generator through the interaction of the stator and rotor magnetic fields. As a result of this transformation, a reaction torque is applied by the generator to the shaft. In the steady state, if we neglect the losses due to friction and windage, this reaction torque is equal and opposite to the prime mover torque and eqn. (3.44) shows that the shaft speed remains constant.

To study the interactions between the mechanical and electrical parts of the system, it is convenient to modify eqn. (3.44) in several ways. First, if we multiply both sides of eqn. (3.44) by the mechanical speed, the right-hand side will be expressed in terms of power rather than torque:

$$J\omega_m\frac{d^2\theta_m}{dt^2} = P_m - P_e \qquad (3.45)$$

Secondly, since we are concerned about deviations between the actual speed and the synchronous speed, it is convenient to measure the angular position of the rotor with respect to a synchronously rotating reference frame. We define

$$\theta_m = \omega_m^{syn} t + \delta_m \tag{3.46}$$

Since ω_m^{syn} is a constant, we can rewrite eqn. (3.45) as

$$J\omega_m \frac{d^2\delta_m}{dt^2} = P_m - P_e \tag{3.47}$$

Finally, electrical and mechanical angles are related by the number of poles in the generator windings,

$$\delta = \frac{N}{2}\delta_m \tag{3.48}$$

Replacing in eqn. (3.47), we obtain

$$\frac{2}{N}J\omega_m \frac{d^2\delta}{dt^2} = P_m - P_e \tag{3.49}$$

To normalise this equation, both sides are usually divided by the MVA or kVA rating of the generator S_B:

$$\frac{2}{N}\frac{J}{S_B}\omega_m \frac{d^2\delta}{dt^2} = P_m^{pu} - P_e^{pu} \tag{3.50}$$

The moment of inertia is then expressed in terms of the inertia constant of the generating unit, which is defined as the ratio of the kinetic energy stored at synchronous speed to the generator kVA or MVA rating,

$$H = \frac{\frac{1}{2}J\omega_m^{syn^2}}{S_B} \tag{3.51}$$

Extracting J from eqn. (3.51) and replacing in eqn. (3.50), we get

$$\frac{2}{N}\frac{2H\omega_m}{\omega_m^{syn^2}}\frac{d^2\delta}{dt^2} = P_m^{pu} - P_e^{pu} \tag{3.52}$$

Normalising the angular frequencies using

$$\omega^{syn} = \frac{N}{2}\omega_m^{syn} \tag{3.53}$$

and

$$\omega^{pu} = \frac{\omega_m}{\omega_m^{syn}} \tag{3.54}$$

we obtain the final version of what is often called the *swing equation*,

$$\frac{2H}{\omega_{syn}}\omega^{pu}\frac{d^2\delta}{dt^2} = P_m^{pu} - P_e^{pu} \tag{3.55}$$

3.5.2 Power transfer in a two-bus system

Consider the simple circuit shown on Figure 3.25. The active and reactive powers injected by the ideal voltage source $\overline{E}$ into the system are given by:

$$P_e = \text{Re}(\overline{S}) = \text{Re}(\overline{E}\ \overline{I^*})$$
$$Q_e = \text{Im}(\overline{S}) = \text{Im}(\overline{E}\ \overline{I^*})$$

(3.56)

The current $\overline{I}$ can be found using Kirchhoff's voltage law,

$$\overline{I} = \frac{\overline{E} - \overline{V}}{jX} = \frac{E\angle\delta - V\angle 0°}{jX} = \frac{E\angle(\delta - 90°) - V\angle(-90°)}{X}$$

(3.57)

Substituting eqn. (3.57) into eqn. (3.56), we obtain

$$P_e = \text{Re}\left[\frac{E^2\angle 90° - EV\angle(\delta + 90°)}{X}\right] = -\frac{EV}{X}\cos(\delta + 90°)$$

$$Q_e = \text{Im}\left[\frac{E^2\angle 90° - EV\angle(\delta + 90°)}{X}\right] = \frac{E^2}{X} + \frac{EV}{X}\sin(\delta + 90°)$$

(3.58)

which simplifies to:

$$P_e = \frac{EV}{X}\sin\delta$$

$$Q_e = \frac{E^2}{X} - \frac{EV}{X}\cos\delta$$

(3.59)

Consider now the simple two-bus power system shown in Figure 3.26. A single generator at bus A is connected to the rest of the power system by two parallel lines. Bus B is represented as connected to a brick wall to suggest that it is part of a very large power system. This system is assumed to have a much larger inertia than the generator connected at bus A. Furthermore, its Thevenin impedance as seen from bus B is assumed to be much smaller than the impedances of the lines connecting A and B and the internal impedance of generator G. Consequently, the magnitude and angle of the voltage at bus B can be assumed to be

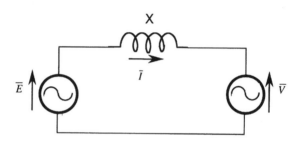

Figure 3.25   Two-source network

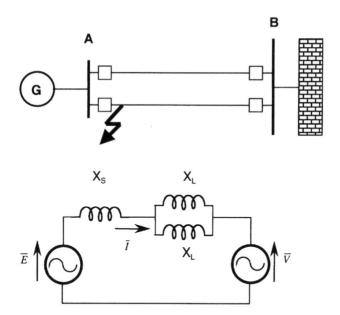

Figure 3.26   Two-bus power system for stability studies and its equivalent circuit

constant, i.e. unaffected by what happens to the left of bus B. These assumptions are traditionally summarised by calling bus B an *infinite bus*.

Figure 3.26 also shows the equivalent circuit of this simple network. As discussed above, the power system beyond bus B can be represented by an ideal voltage source $\overline{V} = V\angle 0°$, which will be taken as the reference for the angles. Generator G is represented by an ideal voltage source $\overline{E} = E\angle\delta$ in series with a reactance X_S. The two transmission lines are assumed to have purely reactive impedances jX_L. Eqn. (3.59) is directly applicable to the system of Figure 3.26 if these various impedances are combined into a single equivalent impedance,

$$X_0 = X_S + \tfrac{1}{2}X_L \tag{3.60}$$

Figure 3.27 shows how the power injected by the generator into the network varies as a function of the angle δ for given values of E and V. The most important conclusion to be drawn from this graph is that the amount of power transferred increases with the angle δ until this angle reaches 90 degrees, where it reaches a maximum value

$$P_{max} = \frac{EV}{X_0}$$

If we neglect the losses in the generator, the mechanical power supplied

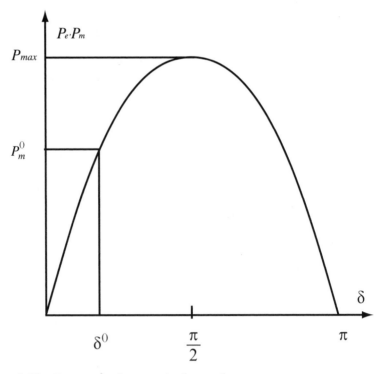

Figure 3.27   Power transfer curve in the two-bus power system

by the prime mover is equal (in the steady state) to the electrical power injected by the generator in the network,

$$P^0_m = P^0_e \tag{3.61}$$

Eqn. (3.59) or Figure 3.27 can thus be used to find the steady-state angle δ° at which the generator operates given the mechanical power.

The reader will undoubtedly have noticed that δ has been used to represent both the mechanical oscillations of the rotor and the angle of the phasor $\overline{E}$ representing the internal emf of the generator. This choice is deliberate, as these two angles are identical. The internal emf $\overline{E}$ is induced in the stator windings of the synchronous generator by the rotation of the rotor flux created by the field winding. It is therefore tied to the position of this rotor. Changes in electrical and mechanical angles are therefore rigidly linked. Any deviation in the position of the rotor alters the amount of electrical power produced by the generator. Conversely, any change in the electrical power flows causes mechanical transients.

3.5.3 *Electro-mechanical transients following a fault*

Consider again the two-bus system shown in Figure 3.26. Suppose that a fault occurs on one of the lines connecting buses A and B, very close to bus A. To analyse the dynamic behaviour of this system, we must consider three periods: *before* the fault, *during* the fault and *after* the fault.

Before the fault, generator G operates in the steady state and injects an electrical power P_e^0 into the network. Since the mechanical power P_m^0 delivered by the prime mover is equal to the electrical power injected in the network, eqn. (3.55) shows that the rotor angle δ^0 is constant.

During the fault, the voltage at bus A is essentially equal to zero. Eqn. (3.59) shows that the active power injected by the generator into the system is therefore also equal to zero. There is thus no longer a balance between the accelerating power provided by the prime mover and the decelerating power resulting from the injection of electrical power into the network. The rotor of the generator accelerates and eqn. (3.55) shows that the rotor angle δ increases.

After an interval of time called the clearing time, the protection system detects the faults and triggers the opening of the breakers at both ends of the faulted line, clearing the fault. At that point, the voltage at bus A is restored to a non-zero value and the network is again capable of transmitting the active power produced by the generator. Since the generator is again transforming mechanical power into electrical power, it applies a decelerating torque to the shaft. If this decelerating torque is applied soon enough, it will succeed in slowing and reversing the increase in the rotor angle δ. Stability will have been maintained. Otherwise, this angle will continue to increase uncontrollably until the protection system of the generator opens the generator breaker to prevent damage to the plant. Stability has been lost.

The purpose of stability studies is thus to determine whether all credible faults will be cleared quickly enough to maintain stability. If the clearing time is given, these studies can be used to determine the maximum load that the network can handle without causing instability in the event of a fault.

3.5.4 *The equal area criterion*

The equal area criterion is a simple graphical method for determining whether a one-machine infinite bus system will remain stable. It provides a useful representation of the factors that affect stability. In practical systems, it may also be used to obtain a first approximation of the stability limit.

This criterion is based on the application of the power transfer curves and is illustrated by Figure 3.28. Before the fault, the generator operates along the power transfer curve defined by the generator internal emf E, the infinite bus voltage V and the equivalent impedance X_0 given by eqn. (3.60). The generator angle and the mechanical power input are related by

$$P^0_m = \frac{EV}{X_0} \sin \delta^\circ \qquad (3.62)$$

During the fault, there is no decelerating electrical torque and the rotor accelerates, increasing the angle δ. By the time the fault is cleared at t^{clear}, this angle has reached the value δ^{clear}. The kinetic energy stored by the rotor is proportional to the area labelled A_1 in Figure 3.28. After the fault clears, the voltage at bus A is restored, and power can again be transmitted between the generator and the infinite bus. However, this transmission takes place on a new power transfer curve because the equivalent impedance between the internal emf and the infinite bus is now

$$X_1 = X_S + X_L \qquad (3.63)$$

because one of the transmission lines has been taken out of service. Since $X_1 > X_0$, this postclearing fault is below the prefault curve. Note that for $\delta = \delta^{clear}$, the electrical power transferred on this new curve is larger than the mechanical power applied by the prime mover. The decelerating

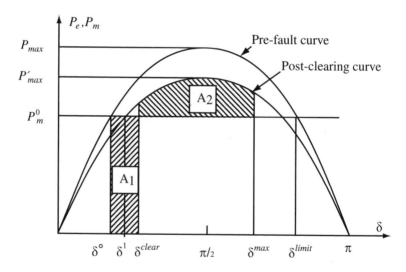

Figure 3.28   Equal area criterion

torque is thus larger than the accelerating torque and the rotor starts to slow down. However, the rotor angle δ continues to increase for a while because of the kinetic energy stored during the fault. The equal area criterion states that δ will increase to a value δ^{max} such that the area A_1 is equal to the area A_2. This criterion is demonstrated in the Appendix (Section 3.8).

Figure 3.28 shows that, if δ reaches δ^{limit} before the equal area criterion is satisfied, the electrical power absorbed by the network becomes smaller than the mechanical power provided by the prime mover. Eqn. (3.55) shows that the rate of increase in δ will then again be positive and stability will be irretrievably lost. On the other hand, if δ^{max} is less than δ^{limit}, δ starts decreasing and stability is maintained. If the damping caused by the electrical and mechanical losses was taken into account, it could be shown that δ settles, after damped oscillations, at the value δ^1.

The equal area criterion suggests that the stability of the system is enhanced if the area A_1 is reduced or if the potential area A_2 is increased. This can be achieved in several ways:

1 *Reducing δ^{clear}* decreases the amount of kinetic energy stored during the fault and provides a larger angular margin ($\delta^{limit} - \delta^{clear}$). A smaller δ^{clear} requires a shorter clearing time (i.e. a faster protection system and faster breakers) or a larger generator inertia.
2 *Reducing P_m^0* also decreases the amount of kinetic energy imparted to the generator during the fault. It also increases the potential energy that the system can absorb after the clearing. Unfortunately, reducing $P_m{}^0$ implies putting a limit on the amount of electrical power supplied by the generator.
3 *Reducing the system impedance* decreases the prefault angle δ^0, increases the angular margin ($\delta^{limit} - \delta^{clear}$) and increases the amount of potential energy that the system can absorb after the clearing. A reduction in impedance can be achieved by connecting the generator to the rest of the system through a line operating at a higher nominal voltage.
4 *Operating at higher voltages* has, in the steady state, the effect of reducing the prefault angle δ^0. Boosting the excitation of the generator to increase the internal emf E after the fault clearing also increases the amount of potential energy that the system can absorb.

3.5.5 Stability studies in larger systems

While the analysis presented above provides useful insights into the mechanisms leading to transient instability in power systems, it relies on a series of simplifying assumptions that may not be justified in an actual power system:

1 Modelling the system as a single generator connected to an infinite bus may not be acceptable if several embedded generators are connected to a relatively

weak network. It may be necessary to model the dynamic interactions between these generators or between one of these generators and large rotating loads.

2 The generator was modelled as a constant voltage behind a single reactance. This very simple model does not reflect the complexity of the dynamics of synchronous generators and may give a misleading measure of the system stability. In particular, it neglects the stabilising effect of the generator's excitation system.

3 Since electrical and mechanical losses have been neglected, the system has no damping. This approximation distorts the oscillations taking place after a fault and overstates the risk of instability.

4 Finally, it may be necessary to model the effect of faults at different locations in the network rather than simply at the terminals of the generator.

Removing these limitations requires the use of much more complex models for the generators, for the network and for the controllers. Unfortunately, the equal area criterion no longer holds under these conditions and another stability assessment method must be implemented.

Most commercial-grade stability assessment programs rely on the numerical solution of the differential and algebraic equations describing the power system. The solution of these equations tracks the evolution of the system following a disturbance. If it shows the rotor angle of one or more generators drifting away from the angles of the rest of the generators, the system is deemed unstable. On the other hand, if all variables settle into an acceptable new steady state, the system is considered stable.

The model of each generating unit contains between two and eight non-linear differential equations, not counting those required for modelling the excitation system. The differential equations of all the generators are coupled through the algebraic equations describing the network. This coupling requires the solution of a power flow at each time step of the solution of the differential equations.

3.5.6 Stability of induction generators

Historically, induction generators have not been significant in large power systems and so they have not been represented explicitly in many power system analysis programs. The behaviour of large induction motors can be important in transient stability studies, particularly of industrial systems, e.g. oil facilities, and so induction motor models are usually included in transient stability programs. These can be used to give a representation of induction generators merely by changing the sign of the applied torque. However, the models used for these generators have often been based on a representation of a voltage source behind a transient reactance similar to that used for simple transient modelling of synchronous generators [8]. The main simplifying assumption is that stator electrical transients are neglected and there is no provision for saturation of the magnetic circuits to be included. Such models may not be reliable for operation at elevated voltages or for investigating rapid transients such as those due to faults. More sophisticated models of induction generators are available in electromagnetic transient programs or can be found in advanced textbooks [9, 10].

3.5.7 Application to an embedded generation scheme

To illustrate the concepts discussed in the previous sections, we will consider the small system shown in Figure 3.29. Let us first assume that faults at bus D are cleared in 100 ms and that the generator at bus E is producing 20 MW. Figure 3.30 shows the oscillations in the rotor angle of this generator following such a fault at bus D. The generator initially accelerates and the rotor angle reaches a maximum value of approximately 120 degrees 150 ms after the fault. It is clear from the figure that

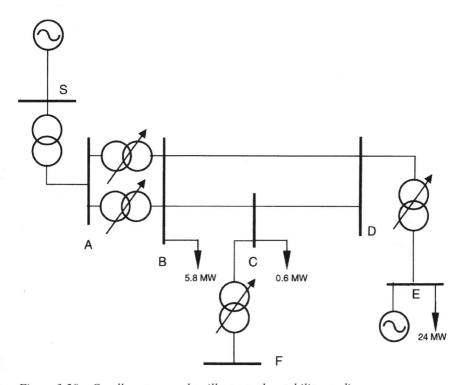

Figure 3.29   Small system used to illustrate the stability studies

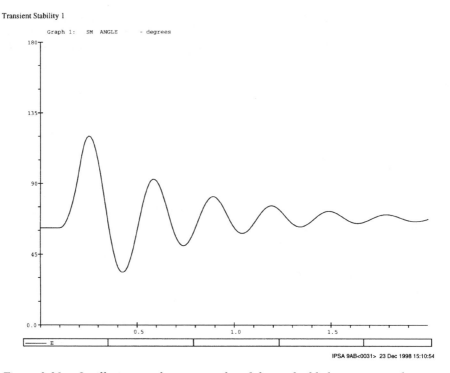

Transient Stability 1

*Figure 3.30   Oscillations in the rotor angle of the embedded generator at bus E
following a fault at bus D when this generator produces 20 MW. The
fault is applied at t = 100 ms and cleared at t = 200 ms*

stability is maintained. Let us now suppose that the protection system
and the switchgear at bus B are such that faults are cleared after only 200
ms. Figure 3.31 shows that stability would be lost even if the generator
was only producing 11.6 MW. 320 ms after the fault, the rotor angle
reaches 180 degrees and 'slips a pole'. At that point, the equipment protecting the generator would immediately take it off-line to protect it from
damage. Stability problems could be avoided by further reducing the
active power output of the generator. Figure 3.32 illustrates the marginally stable case where the generator produces 11.5 MW. The rotor angle
reaches a maximum angle of 165 degrees before dropping. Note that this
angle will ultimately return to its original steady-state value of about 45
degrees but that this may take some time, as the system is marginally
stable. Figure 3.33 illustrates the effect of a fault at bus F, in another 11
kV section of the system. Such faults may occasionally take up to 2 s to
clear. However, since the reactance between the fault and the embedded
generator is large, the system remains stable even for such a long clearing
time. It should be noted that the reactance of the transformer between

Transient Stability 1

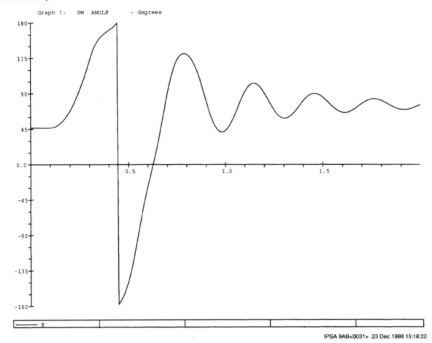

IPSA 9AB<0031> 23 Dec 1998 15:18:22

*Figure 3.31   Oscillations in the rotor angle of the embedded generator at bus E
following a fault at bus D when this generator produces 11.6 MW.
The fault is applied at t = 100 ms and cleared at t = 300 ms. Note
that this particular computer program keeps all angles as being
between +180 degrees and –180 degrees. There is thus no
discontinuity in angle at t = 445 ms – simply a modulus operation*

buses C and F has been adjusted to produce an almost critical case.
Therefore, the rotor angle of the generator does take some time to return
to its prefault steady-state value.

To illustrate the problems associated with voltage instability, the
embedded synchronous generator used in the previous examples has
been replaced by an induction generator. The four graphs of Figure
3.34 show the consequence of a fault at bus D. Prior to the fault, the
voltage at bus D is about 0.98 p.u. and the voltage at bus is E is
maintained at 1.0 p.u. by the action of the tap changing transformer
(graph 1). The induction generator operates at a slip of slightly less
than −1% (graph 2), produces 21 MW (graph 3) and consumes 10.5
MVAr (graph 4). When the fault is applied at $t = 0.5$ s, the voltage at
bus D drops to zero, followed by the voltage at bus E. Since the electrical power produced by the generator also drops to zero, the rotor,

Transient Stability 1

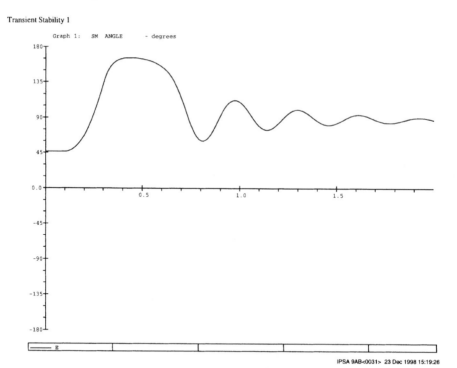

IPSA 9AB<0031> 23 Dec 1998 15:19:26

Figure 3.32   Oscillations in the rotor angle of the embedded generator at bus E following a fault at bus D when this generator produces 11.5 MW. The fault is applied at t = 100 ms and cleared at t = 300 ms

which is still driven by the prime mover, accelerates. As the speed gets larger than the synchronous speed, the slip takes values that are more negative. At t = 2.0 s, the fault is cleared and the voltage at buses D and E recovers. However, due to the very large negative value of the slip, the induction generator absorbs a much larger amount of reactive power. This reactive power is supplied by the transmission system and causes a significant voltage drop in the distribution system. Therefore, the voltage at bus D (and consequently the voltage at bus E) does not fully recover. The induction generator can again inject active power in the system, slowing the increase in slip. However, the slip has already increased so much that the induction generator operates on the back of the torque/speed curve. Since operation on that portion of the curve is unstable, the slip continues to increase. It should be noted that, if this induction generator were connected to a wind turbine, it would have tripped when it reached a speed 10% above synchronous speed to avoid damaging the prime mover.

Transient Stability 1

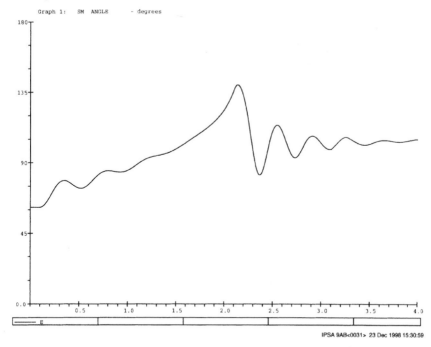

Figure 3.33   *Oscillations in the rotor angle of the embedded generator at bus E*
following a fault at bus F when this generator produces 20 MW. The
fault is applied at t = 0.1 s and cleared at t = 2.1 s

3.6 Electromagnetic transient studies

Power flow, fault and stability calculations are all based on the assumptions that voltages and currents are sinusoidal, that the frequency of
these waveforms is constant, that their amplitudes vary slowly and that
all devices in the system are linear. These assumptions allow us to use
phasor analysis where inductances and capacitances are modelled by
constant reactances and networks are represented by algebraic equations.
The accuracy of the results obtained using phasor analysis is usually very
good, even when studying the effect of faults on system stability. On
the other hand, an accurate modelling of the following components or
phenomena is not possible using phasor analysis:

- Power electronics converters produce voltage and current waveforms
  by switching between various sources. The resulting waveforms are
  definitely non-sinusoidal.
- Magnetic saturation in transformers or rotating machines introduces a
  non-linear relation between voltages and currents.

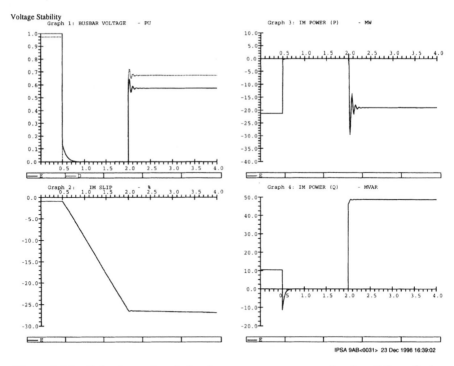

IPSA 9AB<0031> 23 Dec 1998 16:39:02

*Figure 3.34   Behaviour of an induction generator connected at bus E for a fault
at bus D. The fault occurs at t = 0.5 s and is cleared at t = 2.0 s.
Graph 1 shows the voltages at bus D (dashed line) and at bus E
(continuous line). Graph 2 shows the slip of the induction generator.
Graphs 3 and 4 show, respectively, the active and reactive power
output of the generator*

- The operation of circuit breakers and lightning strikes produces fast
  transients that cannot be analysed using a sinusoidal approximation.
  Whenever the effect of such devices and events needs to be modelled
  accurately, we need to revert to the time domain:
- The exact waveform of the voltage and current sources must be considered. This implies simulating the switching of power electronics
  converters or the waveform of a lightning impulse.
- Magnetic devices must be represented by their saturation or hysteresis
  curve.
- Inductances and capacitances must be represented by the differential
  relationship between voltage and current:

$$v_L = L \frac{di_L}{dt}$$

$$i_C = C \frac{dv_C}{dt}$$

(3.64)

Voltage and current waveforms in the system must then be calculated through step-by-step numerical integration of these differential equations.

This type of analysis is usually called electromagnetic transient simulation. Since it requires a considerable amount of computing time, it is usually applied only to the portion of the power system that requires such detailed modelling.

3.7 References

1 GRAINGER, J.J., and STEVENSON, W.D.: 'Power system analysis' (McGraw Hill International Editions, New York, 1994)
2 GLOVER, J.D., and SARMA, M.: 'Power system analysis and design' (PWS Publishing Company, Boston, 1994)
3 WEEDY, B.M., and CORY, B.J.: 'Electric power systems' (John Wiley & Sons, 1998, 4th edn.)
4 KUNDUR, P.: 'Power system stability and control' (McGraw Hill, New York, 1994)
5 ANDERSON, P.M.: 'Analysis of faulted power systems' (IEEE Press, New York, 1995)
6 INTERNATIONAL ELECTROTECHNICAL COMMISSION: 'IEC 909: Short circuit calculation for three-phase ac systems', 1988
7 ELECTRICITY ASSOCIATION: 'Engineering Recommendation G74: Procedure to meet the requirements of IEC 909 for the calculation of short circuit currents in three-phase ac power systems', 1992
8 STAGG, G.W., and EL-ABIAD A.H.: 'Computer methods in power systems analysis' (McGraw-Hill, 1981)
9 KRAUSE, P.C., WASYNCZUK O., and SUDHOFF S.D.: 'Analysis of electric machinery' (IEEE Press, New York, 1994)
10 VAN CUTSEM, T., and VOURNAS, C.: 'Voltage stability of electric power systems' (Kluwer Academic Press, Boston, 1998)

3.8 Appendix: Equal area criterion

The dynamics of the one machine/infinite bus system are governed by the swing equation

$$\frac{2H}{\omega_{syn}} \omega^{pu} \frac{d^2\delta}{dt^2} = P_m^{pu} - P_e^{pu} \tag{3.65}$$

When the system is in the steady state, the generator rotates at nominal speed. While faults and other disturbances cause speed changes that affect the angle δ, these deviations are small compared to the nominal speed. It is thus reasonable to assume that the per unit angular speed ω^{pu} on the left-hand side of eqn. (3.65) is equal to 1.0 p.u., giving

$$\frac{2H}{\omega_{syn}}\frac{d^2\delta}{dt^2} = P_m^{pu} - P_e^{pu} \tag{3.66}$$

Multiplying this equation by $d\delta/dt$ and using

$$\frac{d}{dt}\left(\frac{d\delta}{dt}\right)^2 = 2\left(\frac{d\delta}{dt}\right)\left(\frac{d^2\delta}{dt^2}\right) \tag{3.67}$$

leads to

$$\frac{H}{\omega_{syn}}\frac{d}{dt}\left(\frac{d\delta}{dt}\right)^2 = (P_m^{pu} - P_e^{pu})\frac{d\delta}{dt} \tag{3.68}$$

Multiplying eqn. (3.68) by dt and integrating from δ^0 to an arbitrary angle δ, we get

$$\frac{H}{\omega_{syn}}\int_{\delta^0}^{\delta} d\left(\frac{d\delta}{dt}\right)^2 = \int_{\delta^0}^{\delta}(P_m^{pu} - P_e^{pu})d\delta \tag{3.69}$$

or

$$\frac{H}{\omega_{syn}}\left[\left(\frac{d\delta}{dt}\right)^2\right]_{\delta^0}^{\delta} = \int_{\delta^0}^{\delta}(P_m^{pu} - P_e^{pu})d\delta \tag{3.70}$$

Since for $\delta = \delta^0$, the rotor has not yet accelerated, we have $d\delta/dt = 0$ and

$$\frac{H}{\omega_{syn}}\left(\frac{d\delta}{dt}\right)^2 = \int_{\delta^0}^{\delta}(P_m^{pu} - P_e^{pu})d\delta \tag{3.71}$$

which shows that the change in kinetic energy of the rotor is equal to the integral of the difference between the curves of mechanical and electrical power.

If we choose as the upper limit of the integration δ^{max}, we again have $d\delta/dt = 0$ and eqn. (3.70) gives

$$\int_{\delta^0}^{\delta^{max}}(P_m^{pu} - P_e^{pu})d\delta = 0 \tag{3.72}$$

Separating this integral at $\delta = \delta^{clear}$, we have

$$\int_{\delta^0}^{\delta^{clear}}(P_m^{pu} - P_e^{pu})d\delta + \int_{\delta^{clear}}^{\delta^{max}}(P_m^{pu} - P_e^{pu})d\delta = 0 \tag{3.73}$$

or

$$\underbrace{\int_{\delta^0}^{\delta^{clear}} P_m^{pu}d\delta}_{= A_1} = \underbrace{\int_{\delta^{clear}}^{\delta^{max}}(P_e^{pu} - P_m^{pu})d\delta}_{= A_2} \tag{3.74}$$

which demonstrates the equal area criterion.

A_1 thus represents the amount of kinetic energy that is stored in the rotor during the fault. To maintain stability, this energy must be drawn out of the rotor and stored as potential energy in the network. Subsequent oscillations represent further transformations of this energy between these two forms. These oscillations continue until this energy has been dissipated in electrical and mechanical losses, which are not represented in this model.

Aylesford Co-generation Plant
The plant produces 220 MW heat and 98 MW of electrical energy.

(Source: National Power PLC)

Chapter 4

Generators

Embedded generation plant mainly uses conventional electrical machines or power electronic converter circuits. These are described in detail in a large number of excellent text books, e.g. References 1–5. However, compared to stand-alone generators or even large central power station units, there are significant differences in the way that embedded generators are applied and controlled and some aspects of their behaviour assume increased importance due to their position in the power system or the type of prime mover used. In particular, the presence of rotating machines in the distribution network can significantly alter the flow of fault currents and so needs careful attention.

Figure 4.1 shows an embedded generator connected to a radial distribution circuit. The circuit is connected to the transmission system through one or more tap-changing transformers and has loads connected to it at various locations. This arrangement can easily be studied, in as much detail as required (provided data are available), using the computer based methods discussed in Chapter 3. However, for many initial, but only approximate, calculations the network may be represented by a single equivalent impedance connected to an infinite busbar. Section 3.4.2 and eqn. (3.31) showed that, in the per unit system, the magnitude of this impedance is simply the reciprocal of the short-circuit level (without the generator connected), although the X/R ratio of the network must also be known. The terms short-circuit level and fault level are synonymous. The simple representation of Figure 4.1 is particularly useful as the data required are usually readily available from the distribution utility and it allows a simple impedance to be added in series to the equivalent circuits used to describe the generators. It is used here to help illustrate various important principles.

All large central generators and most stand-alone schemes use synchronous generators for their high efficiency and independent control of real and reactive power. However, some embedded generation plant uses induction generators and there is increasing use of power electronic

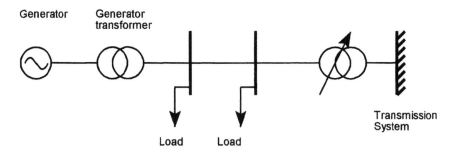

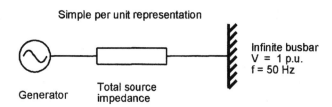

Figure 4.1   Generator connected to a network

interfaces. Induction machines are, of course, widespread on the power system as induction motors but are not widely used as generators, in spite of their apparent simplicity of construction and hence potential economy, because of the defined relationship between real power exported and reactive power drawn. Induction generators do, however, have the significant benefit of providing large damping torques in the prime mover drive train and so are always used in fixed speed wind turbines. Power electronic interfaces are used to connect energy sources which produce DC (e.g. photovoltaics, fuel cells, etc.) but also allow rotating prime movers to operate at variable speed.

4.1 Synchronous generators

Synchronous generators consist of an armature winding located on the stator which is connected to the three phases of the network and a field winding on the rotor which is fed from a source of direct current. The armature winding develops an mmf (magneto motive force) rotating at a speed proportional to the supply frequency. The field winding produces an mmf which is fixed with respect to the rotor. In normal operation the rotor, and hence the field winding, rotates synchronously with the mmf developed by the stator with its relative angle, the load angle, determined by the torque applied to the shaft. The speed of rotation can be reduced at the design stage by increasing the number of pole pairs of the generator as shown in eqn. (4.1).

$$N = f \times 60/p \qquad (4.1)$$

where N is the speed of rotation in rpm, f is the system frequency (50 or 60 Hz) and p is the number of pole pairs.

The stator or armature windings are similar to those found on induction machines but there are a number of different rotor arrangements. Large steam turbine generator sets use turbo-alternators consisting of a cylindrical rotor with a single DC winding to give one pair of poles and hence maximum rotational speed (3000 rpm on 50 Hz systems). Hydro-generators often operate at lower speeds and then use multiple-pole generators with a salient-pole rotor. Smaller engine driven units also generally use salient pole generators. Exceptionally the field winding may be replaced by permanent magnets but this is not commonly found in large generators as, although higher efficiencies can be achieved, direct control of the rotor magnetic field is not possible with this arrangement. Innovative designs of permanent magnet generators are being investigated as they can be constructed with a large number of poles and so operate at low speed and be connected directly to slow speed prime movers, hence removing the requirement for a gearbox in the drive train.

Figure 4.2 shows the simple mechanical analogues of synchronous and

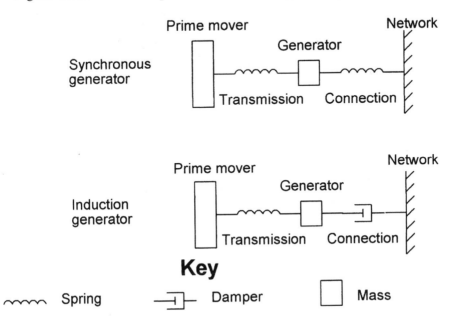

Figure 4.2   Simple mechanical analogues of generators

induction generators. In a synchronous generator, the rotor rotates at constant speed and the rotor angle is a function of the applied torque. Hence the connection of the generator to the network is represented by a spring. In an induction generator, the rotor rotates at a slip speed, slightly faster than the synchronous magnetic field of the stator, and so the rotor speed is a function of the torque applied to the shaft. Hence the connection to the network is represented by a rotational damper. Clearly the simple synchronous machine analogue shown would oscillate perpetually if excited by a torque either due to a disturbance from the network or from the prime mover. In practice this is controlled by damping built into the generator, for example by damper windings on a salient pole generator. These analogues only give a simple picture over a limited linear region of operation but they do allow the operation of the two types of generator to be contrasted.

4.1.1 Steady-state operation

To investigate how a synchronous generator will behave on the power system a simple model is required. Figure 4.3 shows the normal equivalent circuit used to represent steady-state operation, based on the assumptions that: (i) the magnetic circuits are unsaturated, (ii) the air-gap is uniform and any effects of saliency are ignored, (iii) the air-gap flux is sinusoidal, and (iv) the stator resistance is negligible. It may be seen that

$$\overline{V} = \overline{E}_f - j\overline{I}X_s \qquad (4.2)$$

where $\overline{V}$ is the terminal voltage, $\overline{E}_f$ is the internal voltage (function of the field current) and X_s is the synchronous reactance (components due to armature leakage reactance and armature reaction).

For a small embedded generator the terminal voltage is usually held almost constant by the network and so phasor diagrams may be drawn (Figure 4.4) to illustrate the operation of a synchronous generator onto a fixed voltage (or infinite busbar). The power factor of the power delivered to the network is simply cos φ, while the rotor angle (the angle by which the rotor is in advance of the stator field) is given by δ (see

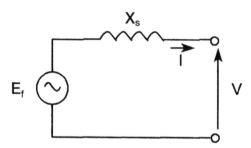

Figure 4.3   Steady-state equivalent circuit of a round rotor synchronous generator

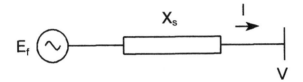

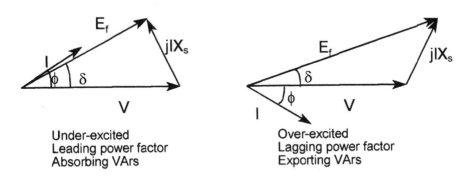

Under-excited
Leading power factor
Absorbing VArs

Over-excited
Lagging power factor
Exporting VArs

Figure 4.4   Phasor diagrams of a round rotor synchronous generator connected to a voltage V

Section 3.5.1 and eqn. 3.48). This is a particular case of the two-busbar system discussed in Chapter 3, and so the active power delivered is simply

$$P = (E_f V/X_s) \sin \delta \qquad (4.3)$$

while the reactive power delivered to the system is

$$Q = E_f V/X_s \cos \delta - V^2/X_s \qquad (4.4)$$

(Note that eqn. (3.59) describes the reactive power flowing into the reactance and so the sign is reversed in eqn. (4.4).)

In normal operation, the rotor angle δ is usually less than 30° and so the cosine term of eqn. (4.4) remains fairly constant. Thus, increasing the torque on the rotor shaft increases the rotor angle (δ) and results in more active power exported to the network. Increasing the field current and hence increasing the magnitude of $\overline{E}_f$ results in export of reactive power. The phasor diagrams of Figure 4.4 show the same rotor angle (and hence the same active power export) but with two different values of excitation:

(a) Underexcited

$$|\overline{E}_f| < |\overline{V}|$$

a leading power factor (using a generator convention and the direction of I as shown), importing reactive power.

(b) Overexcited

$$|\overline{E}_f| > |\overline{V}|$$

a lagging power factor (using a generator convention and the direction of I as shown), exporting reactive power.

It may be noted that if the direction of the *definition* of the current I is reversed, and the machine considered as a motor rather than a generator, then an underexcited motor has a lagging power factor and an overexcited motor has a leading power factor. Of course, if torque is still applied to the shaft then active power will be exported to the network, and if $|\overline{E}_f| > |\overline{V}|$ then reactive power will still be exported irrespective of whether the same machine is called a motor or generator. Therefore, it is often helpful to consider export/import of real and reactive power rather than leading/lagging power factors which rely on the *definition* of the direction of the current flow.

Values of X_s are available from generator manufacturers and some typical ranges are quoted in References 3 and 4. References 1 and 2 show how the analysis can be extended to include the effects of saliency in the rotor and saturation of the magnetic circuits. In practice good power systems analysis programs will include these effects if sufficient data are provided.

The operating chart of a synchronous generator is formed directly from the phasor diagram of Figure 4.4. The phasor diagram is simply scaled by multiplying by V/X_s, which is a constant, to give the phasor diagram of Figure 4.5. The locus of the new phasor VI then describes the operation of the generator. Various limits are applied to account for: (i) the maximum power available from the prime mover, (ii) the maximum current rating of the stator, (iii) the maximum excitation and (iv) the minimum excitation for stability and/or stator end winding heating. These limits then form the boundaries of the region within which a synchronous generator may operate. In practice there may be additional limits including a minimum power requirement and the effect of the reactance of the generator transformer.

The operating chart illustrates that a synchronous generator connected to an infinite busbar of fixed voltage and frequency has essentially independent control over real and reactive power. Real power is varied by adjusting the torque on the generator shaft and hence the rotor angle, while reactive power is adjusted by varying the field current and hence the magnitude of E_f. For example, at point (x) both real and reactive power are exported to the network, at (y) rather more real power is being exported at unity power factor, while at (z) real power is exported and reactive power imported.

Figure 4.6 shows a notional 5 MW synchronous embedded generator driven by a small steam turbine. If the short-circuit level at the point of connection (C) is, say, 100 MVA, with an X/R ratio of, say, 10, then the total source impedance on a 100 MVA base will be approximately

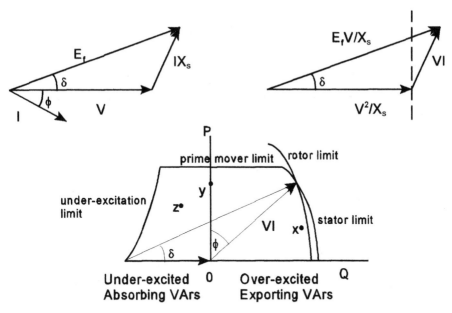

Figure 4.5 *Operating chart of a round rotor synchronous generator connected to an infinite busbar*

$$\overline{Z} = 0.1 + j1.0$$

and with a realistic value of X_s of 1.5 per unit on rating, then, again on a 100 MVA base,

$$X_s = j30$$

Thus it can be seen immediately that $|X_s| \gg |Z|$ and, to a first approximation, the synchronous generator will have a very small effect on network voltage. As a small generator cannot affect the frequency of a large interconnected power system, then the embedded generator can be considered to be connected directly to the infinite busbar. Figure 4.6 is an oversimplification in one important respect in that the other loads on the network are not shown explicitly and these will alter the voltage at the point of connection of the generator considerably. In some power systems, changes in total system load or outages on the bulk generation system will also cause significant changes in frequency.

The conventional method of controlling the output power of a generating unit is to set up the governor on a droop characteristic. This is shown in Figure 4.7, where the line (a–b) shows the variation in frequency (typically 4%) required to give a change in the power output of the prime mover from no-load to full-load. Thus with 1 per unit frequency (50Hz) the set will produce power P1. If the frequency falls by 1% the output power increases to P3, while if the system frequency rises by

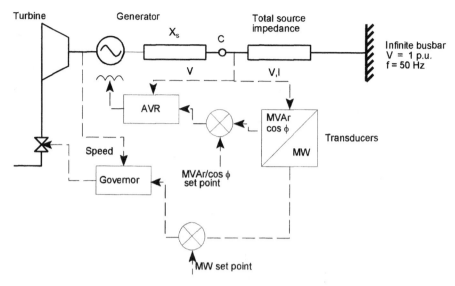

Figure 4.6   Control of an embedded generator

1% the output power is reduced to P2. This, of course, is precisely the behaviour required from a generator which can influence the system frequency; if the frequency drops more power is required, while if the frequency rises less power is needed. The position of the droop line can be changed using the 'speeder gear' and so by moving the characteristic to (a′–b′) the power output can be restored to P1 even with an increased system frequency or by moving to (a″–b″) for a reduced system frequency.

A similar characteristic can be set up for voltage control (Figure 4.8) with the axes replaced by reactive power and voltage. Again, consider the droop line (a–b). At 1 per unit voltage no reactive power is exchanged with the system (operating point Q1). If the network voltage rises by 1% then the operating point moves to Q2 and reactive power is imported by the generator, in an attempt to control the voltage rise. Similarly if the network voltage drops the operating point moves to Q3 and reactive power is exported to the system. Translating the droop lines to (a′–b′) or (a″–b″) allows the control to be reset for different conditions of the network. The slope of both the frequency and voltage droop characteristics can also be changed if required.

These conventional control schemes may not be appropriate for small embedded synchronous generators. For example, an industrial CHP plant may wish to operate at a fixed power output, or fixed power exchange with the network, irrespective of system frequency. Similarly, operation with no reactive power exchange with the network may be desirable to minimise reactive power charges. If the generators are oper-

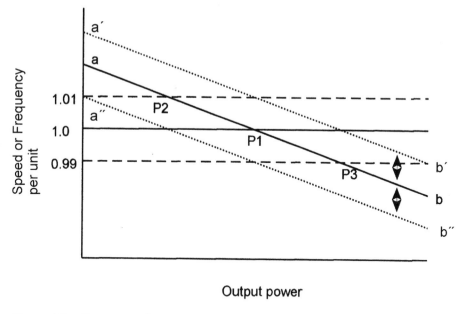

Figure 4.7 Conventional governor droop characteristic

ated on the simple droop characteristics illustrated in Figures 4.7 and 4.8 then both real and reactive power outputs of the generator will change constantly with network conditions, as the network voltage and frequency vary under external influences.

Therefore, for many relatively small embedded generators on strong networks, control is based on real and reactive power output rather than on frequency and voltage, as might be expected in stand-alone installations or for large generators. As shown in Figure 4.6, voltage and current signals are obtained at the terminals of the generator and passed to transducers to measure the generated real and reactive power output. The main control variables are MW, for real power, and MVAR or cos φ for reactive power. A voltage measurement is also supplied to the automatic voltage regulator (AVR) and a speed/frequency measurement to the governor but, in this mode of control, these are supplementary signals only. It may be found convenient to use the MW and MVAr/cos φ error signals indirectly to translate the droop lines and so maintain some of the benefits of the droop characteristic, at least during network disturbances, but this depends on the internal structure of the AVR and governor.

However, the principal method of control is that, for real power control, the measured (MW) value is compared to a set point and then the error signal fed to the governor which, in turn, controls the steam supply to the turbine. In a similar manner the generator excitation is controlled to either an MVAr or cos φ setting. The measured variable is compared

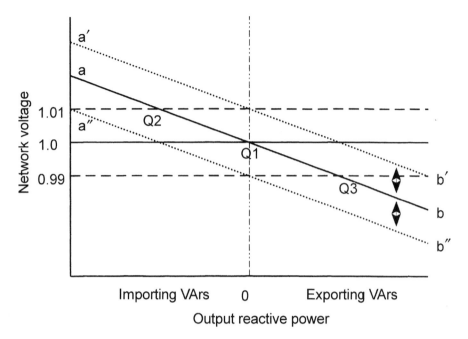

Figure 4.8   Quadrature droop chacteristic for generator excitation control

to a set point and the error passed to the AVR and exciter. The exciter
then controls the field current and hence the reactive power output.

It may be noted that the control scheme shown in Figure 4.6 pays no
attention to the conditions on the power system. The real power output is
controlled to a set point irrespective of the frequency of the system while
the reactive power is controlled to a particular MVAr value or power
factor irrespective of network voltage. Clearly for relatively large embedded generators which can have an impact on the network this is unsatis-
factory and more conventional control schemes such as voltage control
with quadrature droop and the use of supplementary frequency signals
to improve the governor response are likely to be appropriate [6]. These
are well established techniques used wherever a generator has a significant impact on the power system but there remains the issue of how to
influence the owners/operators of embedded generation plant to apply
them. Operating at non-unity power factor increases the electrical losses
in the generator while varying real power output in response to network
frequency will have implications for the prime mover and steam supply if
it is operated as a CHP plant. Moreover, as increasing numbers of small
embedded generators are connected to the network it will become
important to co-ordinate their response both to steady-state network
conditions and to disturbances. At present, in the UK at least, the technical and commercial/administrative arrangements for the co-ordination

of the operation of large numbers of small (<50 MW), independently owned embedded generators are not yet in place.

4.1.2 Excitation systems

The performance of a synchronous generator is strongly influenced by its excitation system particularly with respect to transient and dynamic stability, and the ability of the generator to deliver sustained fault current. Supply of sustained fault current is particularly important for embedded generators due to their relatively small ratings and the long clearance times typically found in distribution protection systems. A small embedded generator, with the same per unit machine parameters on rating as a large generator, will provide fault current only in proportion to the machine ratings. Further, distribution networks are often protected with time-delayed overcurrent protection which, because of the way it is set (or graded) can require fault currents considerably higher than the circuit continuous rating in order to operate quickly. Thus, the ability of a small generator to provide adequate fault current requires careful attention during the design of the embedded generation scheme.

On some older generators, a DC generator, with a commutator, was used to provide the field current, which was then fed to the main field via slip rings on the rotor. Equipment of this type can still be found in service, although often with modern AVRs replacing the rather simple voltage regulators which were used to control the field of the DC generator and hence the main excitation current. However, more modern excitation systems are generally of two types: brushless or static.

Figure 4.9 is a schematic representation of a brushless excitation

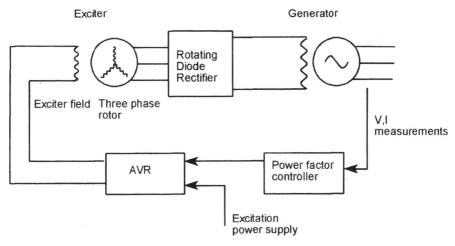

Figure 4.9   Brushless excitation system

system. The exciter is simply an alternator, much smaller than the main generator and with a stationary field and a rotating armature. A full wave diode bridge is mounted on the rotating shaft to rectify the three-phase output of the exciter rotor to DC for the field of the main generator. The exciter field is controlled by the AVR, which, as discussed earlier, is itself controlled by the power factor controller to allow control either to a constant power factor or to a defined reactive power output. Power for the exciter may be taken either from the terminals of the main generator (self-excited) or from a permanent magnet generator (separately excited). The permanent magnet generator is mounted on an extension of the main generator shaft and continues to supply power as long as the generator is rotating. In contrast, a simple self-excited scheme may fail to operate correctly if a close-up fault reduces the voltage at the terminals of the generator as the generator cannot then provide fault current just when it is needed. This is of particular concern for embedded generators connected directly to the distribution network as the impedance of a generator transformer will tend to help maintain the generator terminal voltage in the event of a network fault. It is possible to use current transformers as well as a voltage transformer to supply excitation power in a separately excited scheme but, if costs allow, a more reliable method is to use a permanent magnet pilot exciter.

Figure 4.10 shows a static excitation system in which controlled DC is

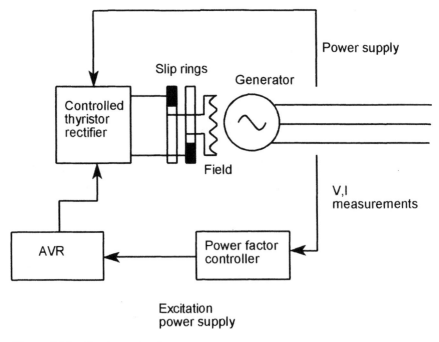

Figure 4.10   Static excitation system

supplied by a controlled thyristor rectifier and fed to the generator field via slip rings. The power supply for the thyristor rectifier is taken from the terminals of the generator. The main advantage of the static exciter is improved response as the field current is controlled directly by the thyristor rectifier but, of course, if the generator terminal voltage is depressed too low then excitation power will be lost. It is again possible to provide the power supply from both voltage and current transformers at the generator terminals but it is doubtful whether the improved response of the static exciter over a permanent magnet brushless scheme can be justified for many small embedded generators.

In addition to the main types described, there are a large number of innovative designs of excitation systems which have been developed over the years particularly for smaller generators. These include the use of magnetic circuits for the no-load excitation and current transformer compounding for the additional excitation required as current is drawn. Although such techniques may work robustly on stand-alone systems they are almost impossible to model for studies of embedded generation schemes. For larger generators and their excitation systems the manufacturers are usually able to supply the so-called IEEE exciter models. These refer to the structure of excitation system models which have been developed by the IEEE [7] and are included in most power system analysis programs.

4.1.3 Operation during network disturbances

The excitation and governor systems of a generator can have a significant impact on its performance during network disturbances. The ability of a synchronous generator to remain stable during network faults (transient stability) is discussed in Chapter 3 and can be simulated provided data are available. However, eqn. (4.3) can be rewritten for transient conditions, to illustrate the potential benefit of fast-acting excitation, as:

$$P = (E_f' V/X') \sin \delta \qquad (4.5)$$

where E_f' is the transient internal voltage of the generator and X' is the transient reactance.

The transient reactance is determined by the generator design but the transient internal voltage (E_f') can be increased by a fast-acting excitation system. Thus, by using a high response excitation system, the power transfer capability of the generator can be maintained, even if the network voltage (V) is depressed by a fault. Hence the generator can remain stable for longer clearing times, or at higher loading conditions, than would be possible with a slower AVR/exciter.

The provision of tight, fast-response control over the generator terminal voltage unfortunately has the effect of reducing generator damping and can, in some circumstances, lead to oscillatory instability. This is unlikely to be an issue for well designed small embedded generation schemes connected to a high short-circuit level but may need to be considered when a large embedded generator exports to an electrically weak network. The tuning of AVRs and the possible application of power system stabilisers is a topic well beyond the scope of this book, and requires specialist advice.

The contribution of a round rotor synchronous generator to a three-phase fault is usually described by an expression of the form

$$I = E_f[1/X_d + (1/X_d' - 1/X_d)\, e^{-t/T_d'} + (1/X_d'' - 1/X_d')\, e^{-t/T_d''}] \cos(\omega t + \lambda)$$
$$-E_f(1/X_d'')\, e^{-t/T_a} \cos \lambda \qquad (4.6)$$

where

X_d = direct axis synchronous reactance
X_d' = direct axis transient reactance
X_d'' = direct axis subtransient reactance
E_f = prefault internal voltage
T_d' = direct axis transient short-circuit time constant
T_d'' = direct axis subtransient short-circuit time constant
T_a = armature (DC) time constant
λ = angle of the phase at time zero
ω = system angular velocity.

Note: Eqn. (4.6) is written in the conventional form based on two-axis electrical machine theory [1–4] . For a round rotor machine the direct axis (i.e. the axis of the field winding) has the same value of reactances as the quadrature axis. The subtransient, transient and synchronous reactances are used to represent the performance of the machine at different times after the fault defined by the corresponding time constants. The armature (DC) time constant is used to describe the decay of the DC offset of the fault current.

The last term of eqn. (4.6) describes the DC component which depends on the point-on-wave at which the fault occurs, while the remainder describes the fundamental frequency AC component. The short-circuit time constants (T_d' and T_d'') and the armature time constant (T_a) are not fixed values but depend on the location of the fault. In particular,

$$T_a = (X_d'' + X_e)/\omega(R_a + R_e) \qquad (4.7)$$

where X_e is the external reactance (to the fault), R_e is the external resistance (to the fault) and R_a is the armature resistance.

Synchronous machine impedances have X/R ratios which are much larger than those of distribution circuits. Hence a fault close to a synchronous generator will have an armature time constant (T_a) and hence a DC component which is much longer than for a remote fault. This is an important consideration for embedded generation schemes. Traditional, passive distribution systems, fed from HV networks through a succession of transformers, can be considered to have fault currents with a very rapid decay of the DC offset and an essentially constant AC component. In contrast, faults close to generators, or large motors, will have slower decaying DC offset and decaying AC components. This is recognised in IEC 909 [8], which recommends two different calculation approaches for these two situations, i.e. the 'far-from-generator-short-circuit' and the 'near-to-generator-short-circuit' . Engineering Recommendation G74 [9] also discusses the various computer based modelling approaches which may be used to represent this effect.

Figure 4.11 shows a simulation of the response of a synchronous generator to a close-up and remote fault. (For clarity, only the phase with maximum offset is shown.) Obviously the fault current is much larger for the close-up fault than for the remote fault. The close-up fault shows the longer decay in the DC components and the reducing AC component. The remote fault shows the very rapid decay of the DC component and only a very small reduction in the AC current.

Figure 4.12 shows a curve which is typical of that supplied by manufacturers to describe the fault current capability of a small synchronous generator on to a three-phase fault at its terminals. In this case the current is expressed in RMS values with logarithmic scales on the axes. It may be seen that the expected decay occurs up to, say, 200 ms but then the excitation system operates to boost the fault current back up to three times full load output. This is necessary as distribution protection is usually time-graded and so sustained fault current is required if current operated protection is to function effectively. This ability to boost the fault current (or to increase E_f in eqn. (4.6)) depends critically on the excitation scheme which has been chosen. Depending on the generator design a 3 per unit sustained fault current on to a terminal's short circuit may require 'field forcing' to an internal voltage of eight to ten times that needed at no-load.

A paper by Griffith [10] gives an excellent review of various types of generator excitation including a discussion of field forcing and the possible use of voltage-controlled overcurrent protection for small embedded generator schemes.

4.2 Induction generators

An induction generator is, in principle, an induction motor with torque applied to the shaft, although there may be some modifications made to the machine design to optimise its performance as a generator. Hence it consists of an armature winding on the stator and, generally, a squirrel cage rotor. Wound rotor induction machines are used in some specialised embedded generating units, but these are not common, while squirrel

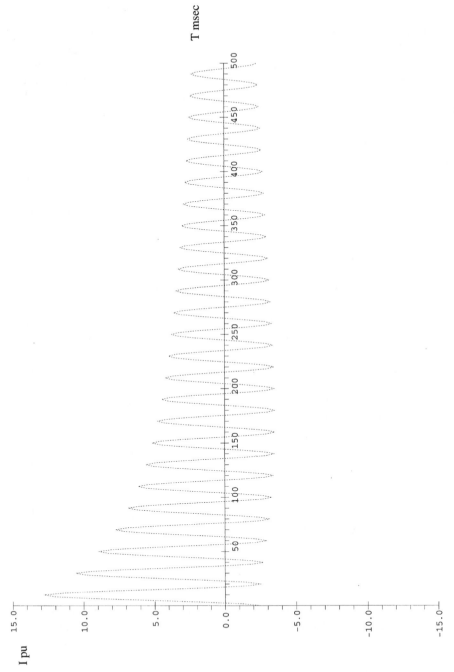

Figure 4.11a   Fault current of a synchronous generator (phase with maximum offset): Close-up fault

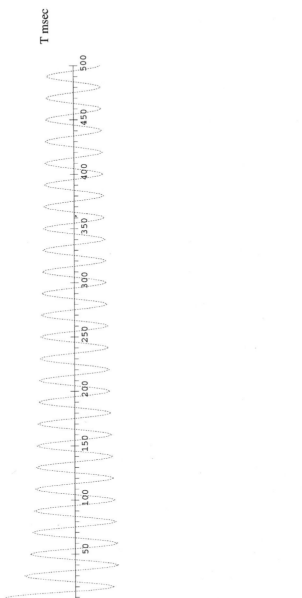

Figure 4.11b   Fault current of a synchronous generator (phase with maximum offset): Remote fault

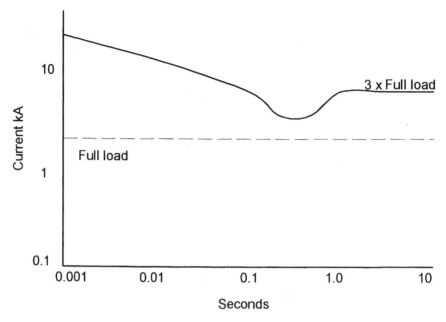

Figure 4.12   Short-circuit decrement curve for a small synchronous generator

cage induction machines are found in a variety of types of small generating plant. The main reason for their use is the damping they provide for the drive train (see Figure 4.2) although additional benefits include the simplicity and robustness of their construction and the lack of requirement for synchronising. The damping is provided by the difference in speed between the rotor and the stator mmf (the slip speed) but as induction generators increase in size their natural slip decreases [11], and so the transient behaviour of large induction generators starts to resemble that of synchronous machines. Induction generators are found in all fixed speed, network connected wind turbines but have also been used in hydro-sets for many years. Reference 12 describes very clearly both the basic theory of induction generators and their application in small hydro-generators in Scotland in the 1950s.

4.2.1 Steady-state operation

The steady-state behaviour of an induction generator may be understood using the familiar (Steinmetz) induction motor equivalent circuit, based on the transformer representation (Figure 4.13). The slip speed (s) is given by

$$s = (\omega_s - \omega_r)/\omega_s \qquad (4.8)$$

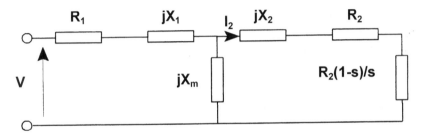

Figure 4.13 Steady-state equivalent circuit of induction machine. Positive phase sequence for balanced operation

R_1 = stator resistance, X_1 = stator leakage reactance, X_m = magnetising reactance, R_2 = rotor resistance (referred to stator), X_2 = rotor leakage reactance (referred to stator), s = slip (negative for generator operation), V = applied voltage (per phase) and I_2 = rotor current

where ω_s is the angular velocity of the stator field and ω_r is the angular velocity of the rotor, which is negative for generator operation.

Working in per unit quantities, the mechanical power of the rotor is given by

$$P_{mech} = I_2^2 R_2 (1 - s)/s \qquad (4.9)$$

while the rotor copper losses are given by

$$P_{losses} = I_2^2 R_2 \qquad (4.10)$$

The usual simple analysis of this circuit relies either on moving the magnetising branch to the supply terminals (the so-called approximate equivalent circuit [1]) or by using a Thevenin transform to eliminate the shunt branch [2].

Considering the approximate equivalent circuit, then the current flowing in the rotor circuit is given simply by

$$I_2 = V/(R_1 + R_2/s + j(X_1 + X_2)) \qquad (4.11)$$

and the total power supplied to the rotor is the sum of the copper losses and the developed mechanical power:

$$P_{rotor} = I_2^2 R_2/s = \{V/(R_1 + R_2/s + j(X_1 + X_2))\}^2 R_2/s \qquad (4.12)$$

This then allows the familiar torque slip curve of an induction machine to be drawn (Figure 4.14). It may be seen that for this example (a 1 MW, 690 V, 4-pole induction generator) the pull-out torques in both the motoring and generating regions are similar at 2 per unit (2 MW). Normal generating operation occurs at up to −1 per unit torque, which corresponds to a slip of −0.7%.

The normal operating locus of an induction machine may also be described in terms of real and reactive power (in a similar manner to the synchronous machine operating chart of Figure 4.5). This is the well

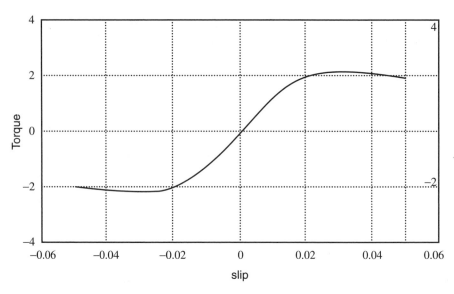

*Figure 4.14   Torque–slip curve of a directly connected 1 MW induction generator,
plotted between −5%/+5% slip*

Applied torque in per unit. Nominal slip = −0.7%(−0.007)

known circle diagram and is shown in Figure 4.15. Compared to a syn-chronous machine, the major difference is that an induction generator can *only* operate on the circular locus and so there is always a defined relationship between real and reactive power. Hence independent control of the power factor of the output of a simple induction generator is not possible. For example, at point B the generator is exporting active power but importing reactive power, while at point A no power is exported but the no-load reactive power is absorbed. It may be seen that the power factor decreases with decreasing load.

Figure 4.16 shows part of the circle diagram of the example 1 MW induction generator and indicates that at no-load some 250 kVAr is drawn as excitation reactive power, rising to over 500 kVAr at the rated output. The no-load real power loss is not represented fully in the approximate equivalent circuit used to develop the diagram.

To improve the power factor it is common to fit local power factor correction (PFC) capacitors at the terminals of the generator. These have the effect of shifting the circle diagram, as seen by the network, down-wards along the *y*-axis. It is conventional to compensate for all or part of the no-load reactive power demand although, as real power is exported, there is additional reactive power drawn from the network.

If a large induction generator, or a number of smaller induction gen-erators, are connected to a network with a low short-circuit level, then the source impedance, including the effect of any generator transformers,

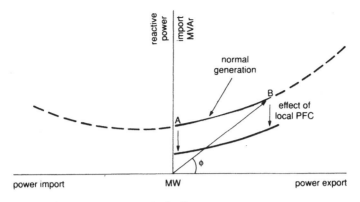

Figure 4.15   Induction generator circle diagram

can become significant. Hence the equivalent circuit can be extended, as shown in Figure 4.17, to include the source impedance in the stator circuit. As an example a group of ten of the 1 MW generators, as might be found in a wind farm, is considered. Each generator is compensated with 200 kVAr of power factor correction capacitors and connected through a 1 MVA generator transformer of 6% reactance to a busbar with a short-circuit level of 100 MVA, which is represented by the source impedance connected to an infinite busbar. The group of ten generators is then considered as an equivalent single 10 MW generator (Figure 4.18). In the per unit system this transformation is achieved conveniently by maintaining all the per unit impedances of the generators, capacitors and transformers constant but merely changing the base MVA of the calculation. This has the effect of increasing the effective impedance of the connection to the infinite busbar by the number of generators (i.e. ten).

Figure 4.19 shows the torque–slip curve of the 10 MW coherent generator. It may be seen that the pull-out torque has dropped significantly to just over 1 per unit (1 MW per turbine) due to the additional source impedance. The slip at which pull-out occurs has not changed but the drop in torque is asymmetric, with the magnitude of the generating pull-out torque maintained rather higher than that when motoring. Increasing the number of turbines in the wind farm effectively increases the impact of the source impedance and would lead to instability when the generator is no longer able to transmit the torque applied by the prime mover. The addition of the power factor correction capacitors has translated the circle diagram towards the origin, as shown in Figure 4.20, but the increased reactive power demand at an active power output of more than 1 per unit can also be seen. Figure 4.21 shows the variation of reactive power drawn from the infinite busbar with slip and it may be

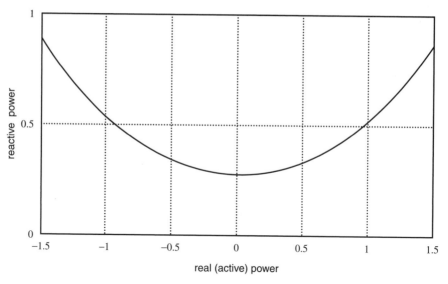

Figure 4.16   Part of circle diagram of 1 MW induction generator
Normal operating region: 0/–1 MW, 250/500 kVAr

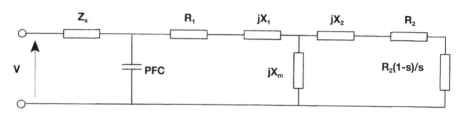

Figure 4.17   Steady-state equivalent circuit of induction machine connected through a source impedance

Power factor correction included
R_1 = stator resistance, X_1 = stator leakage reactance, X_m = magnetising reactance, R_2 = rotor resistance (referred to stator), X_2 = rotor leakage reactance (referred to stator), s = slip (negative for generator operation), V = applied voltage (per phase), Z_s = source impedance and PFC = power factor correction

seen that, in this example, if the 10×1 MW turbines all accelerated beyond their pull-out torque (and remained connected) then some 250 MVAr would be demanded from the network. This would clearly lead to voltage collapse in the network although, in practice, the generators would have tripped on overspeed or undervoltage. This steady-state stability limit might be thought of as analogous to that of a synchronous generator described by eqn. (4.3). If the steady-state stability limit of a synchronous generator is exceeded, the rotor angle increases beyond 90° and pole slipping occurs. With an induction generator, when the pull-out

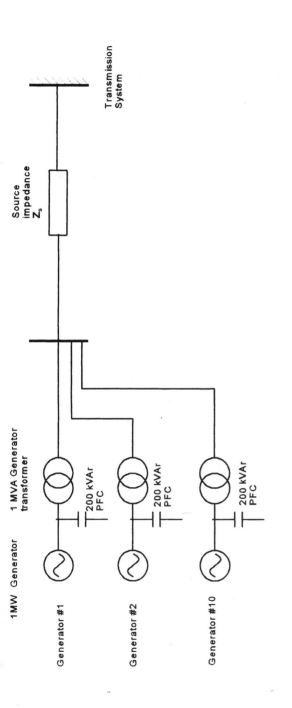

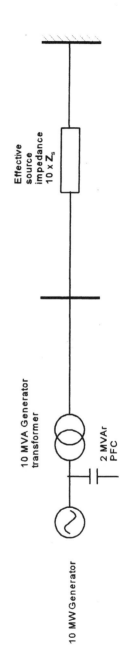

Figure 4.18   Representation of 10×1 MW coherent generators as a single 10 MW generator

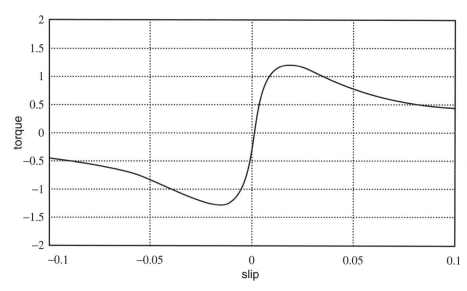

*Figure 4.19   Torque–slip curve of 10 × 1 MW coherent induction generators
connected to a 100 MVA busbar*

torque is exceeded, excess reactive power is drawn to collapse the voltage, and the generator accelerates until the prime mover is tripped.

For large wind farms on weak networks, including large offshore installations which will be connected to sparsely populated coasts, this form of instability may become critical. The voltage change in a radial circuit with an export of active power and import of reactive power is given approximately by

$$\Delta V = (PR - XQ)/V \qquad (4.13)$$

It is frequently found that the circuit X/R ratio happens to be roughly equivalent to the P/Q ratio of an induction generator at near full output. Hence it may occur that the magnitude of the voltage at the generator terminals changes only slightly with load (although the relative angle and network losses will increase significantly) and so this potential instability may not be indicated by abnormal steady-state voltages.

A power-flow program with good induction machine steady-state models will indicate this potential voltage instability and fail to converge if the source impedance is too high for the generation proposed, although it is important to recognise that the governors on some embedded generator prime movers may not be precise and so operation at more than nominal output power must be investigated. The phenomena can be investigated more accurately using an electro-magnetic transient or transient stability program with complete transient induction machine models or using one of the more recently developed continuation power-

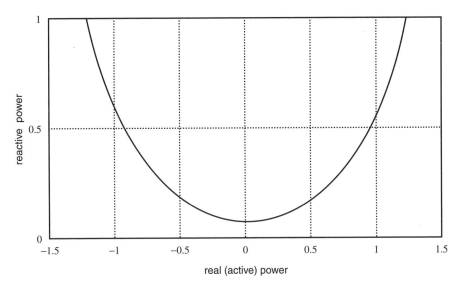

Figure 4.20 *Part of circle diagram of compensated 10 × 1 MW coherent*
induction generators
(Note reduction in no-load reactive power demand but rapid increase at
greater than 1.2 per unit active power)

flow programs which can be used to find the actual point of voltage
instability. A large induction generator, or collection of induction generators, which are close to their steady-state stability limit will, of course,
be more susceptible to transient instability caused by network faults
depressing the voltage.

4.2.2 Connection of an induction generator

A synchronous generator is connected to the network by running it up to
synchronous speed, applying the field voltage and carefully matching the
magnitude and phase of the output voltage of the generator with that of
the network. Only when the generator and network voltages are closely
matched is the circuit breaker closed and the generator connected to the
network.

An isolated induction generator cannot produce a terminal voltage as
there is no source of reactive power to develop the magnetic field. Hence
when an induction generator is connected to the network there is an
initial magnetising inrush transient, similar to that when a transformer
is energised, followed by a transfer of real (and reactive) power to bring
the generator to its operating speed. For a large embedded induction
generator the voltage transients caused by direct-on-line starting are
likely to be unacceptable. Therefore, to control both the magnetising
inrush and subsequent transient power flows to accelerate or decelerate

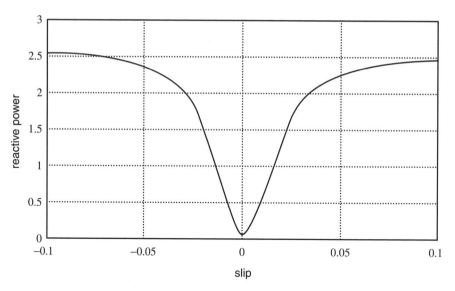

*Figure 4.21   Variation of reactive power drawn from network with slip
(10 × 1 MW coherent induction generators)*

the generator and prime mover it is common to use a 'soft-start' circuit (Figure 4.22). This merely consists of a back-to-back pair of thyristors which are placed in each phase of the generator connection. The soft-start is operated by controlling the firing angle of the thyristors so building up the flux in the generator slowly and then also limiting the current which is required to accelerate the drive train. Once the full voltage has been applied, usually over a period of some seconds, the bypass contactor is closed to eliminate any losses in the thyristors. These soft-start units can be used to connect either stationary or rotating induction generators and, with good control circuits, can limit the magnitudes of the connection currents to only slightly more than full-load current. Similar units are, of course, widely used for starting large induction motors.

4.2.3 Self-excitation

Control over the power factor of the output of an induction generator is only possible by adding external equipment, and the normal method is to add power factor correction capacitors at the terminals, as shown in Figure 4.17. If sufficient capacitors are added then all the reactive power requirement of the generator can be supplied locally and, if connection to the network is lost, then the generator will continue to develop a voltage. In terms of embedded generation plant this is a most undesirable operating condition as, depending on the saturation characteristics of the generator, very large distorted voltages can be developed as the generator

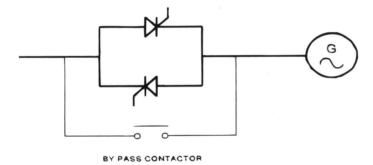

Figure 4.22   Soft-start unit for an induction generator (one phase only shown)

accelerates. This phenomenon of 'self-excitation' has been reported as causing damage to load equipment connected to the isolated part of a network fed from induction generators with power factor correction.

If the connection to the network is lost, the slip is small and, as the stator leakage reactance and resistance are much less than the magnetising reactance, so the equivalent circuit of Figure 4.17 can be reduced to that of Figure 4.23 [1,12]. The magnetising reactance is shown as variable as its value changes with current due to magnetic saturation. Figure 4.23 represents a parallel resonant circuit with its operating points given by the intersection of the reactance characteristics of the capacitors (the straight lines) and the magnetising reactance which saturates at high currents. Thus, at frequency f_1 the circuit will operate at 'a' while as the frequency (or rotational speed of the generator) increases to f_2 the voltage will rise to point 'b'. It may be seen that the voltage rise is limited only by the saturation characteristic of the magnetising reactance. Self-excitation may be avoided by restricting the size of the power factor correction capacitor bank to less than that required to make the circuit resonant at any credible generator speed (frequency), while its effect can be controlled by applying fast-acting overvoltage protection on the induction generator circuit. Most currently available power system analysis programs do not include a representation of saturation in their induction machine models and so cannot be used to investigate this effect. In the detailed models found in electro-magnetic programs, saturation can be included if data are available. However, as self-excitation, and indeed any form of 'islanded' operation, is a condition generally to be avoided, detailed investigation is not necessary for most embedded generation schemes.

4.2.4 Operation during network disturbances

Induction generators have a low impedance to unbalanced voltages and so will draw large currents if the phase voltages of the distribution system are not balanced. This may be seen from Figure 4.24, which shows

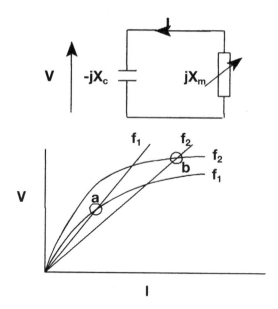

Figure 4.23   Representation of self-excitation of an induction generator [1]

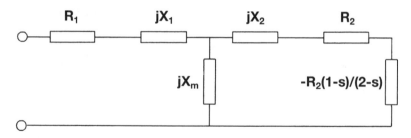

Figure 4.24   Steady-state equivalent circuit of induction machine. Negative phase sequence for unbalanced operation

the negative phase sequence equivalent circuit of an induction machine [13]. In normal operation the slip tends to zero and so the rotor effective resistance is reduced to $R_2/2$. The effect of the unbalanced current is to increase heating in the generator and to impose a torque ripple on the drive train. The phase voltages of rural distribution networks are often unbalanced by the connection of single-phase loads, and a number of embedded generators using induction machines on weak rural networks have experienced difficulty due to excessive unbalanced currents. The effect is often to cause nuisance tripping of the generator phase unbalance protection at particular times when the network load is switched (e.g. when the night storage heater load is connected by time switches). The only solution to this problem is for the utility to try to balance the distribution circuit load across the phases. Rural 11 kV circuits in England and Wales which use single-phase transformers and subcircuits are particularly prone to this problem.

The behaviour of an induction generator under fault conditions is rather different to that of a synchronous generator. A three-phase fault on the network interrupts the supply of reactive power needed to maintain the excitation of the induction generator and so there is no sustained contribution to a symmetrical fault. Figure 4.25 shows one phase of a simulation of a 1 MW, 3.3 kV induction generator. It may be seen that the fault current decays within 100–200 ms.

An expression similar to eqn. (4.6), but with $1/X_d$ omitted, may be used to describe the fault current contribution of an induction generator to a three-phase fault at its terminals. However, owing to difficulties with obtaining data it is often reduced to

$$I = E/X'' \left(\cos \left(\omega t + \lambda \right) e^{-t/T''} + \cos \left(\lambda \right) e^{-t/T_a} \right) \qquad (4.14)$$

where

$T'' = X''/\omega R_{2s}$
$T_a = X''/\omega R_1$
$X'' = X_1 + X_{2s}X_m/(X_{2s} + X_m)$

E is the network voltage (sometimes increased by a factor [9]) and subscript 's' indicates standstill (locked rotor values) of the normal equivalent circuit parameters. As for a synchronous generator, any external impedance involved must be added to the stator impedance.

Unbalanced faults on the network may lead to sustained fault contributions from induction generators and in some cases a rise in current on the unfaulted phase. Appropriate computer simulations are required for an accurate representation of the behaviour of an induction generator feeding a sustained unbalanced fault. The fault current from an induction generator is generally not relied on for the operation of any

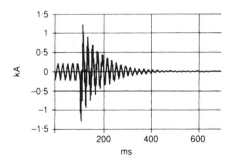

Figure 4.25   *Fault current of induction generator with three-phase fault at its terminals*
Phase shown with minimum DC offset

protective relays. Therefore, when a fault occurs on a distribution system connected to an induction generator, the fault current from the network source is used to operate the distribution system overcurrent protection. This isolates the generator, and so overvoltage, overfrequency or loss-of-mains relays are then used to trip its local circuit breaker and prime mover. This sequential tripping of the generator using voltage, frequency or overspeed protection is necessary as the generator is not capable of providing sustained fault current.

4.2.5 Advanced shunt compensation for induction generators

The use of switched capacitors for power factor correction of induction generators has a number of difficulties. If the degree of compensation approaches the no-load reactive power requirement of the generator then there is the danger of self-excitation while the capacitors can only be switched occasionally and in discrete steps. These limitations can be overcome if the reactive power required is supplied by a power electronic compensator rather than by fixed capacitors. This idea was investigated in a CEU supported research project and an 8 MVAr STATCOM was installed on a 24 MW wind farm in Denmark, Figure 4.26 [14]. The wind farm consisted of 40×600 kW fixed-speed wind turbines connected to a 15 kV power collection system. The wind farm was then

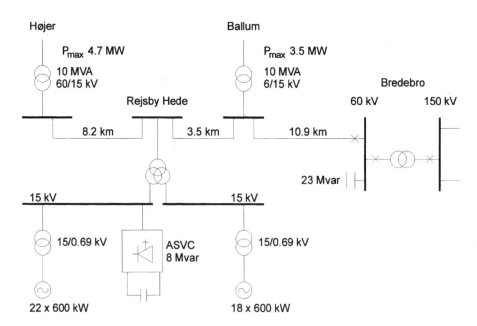

Figure 4.26 Schematic diagram of the Rejsby Hede wind farm, STATCOM (ASVC) and distribution network

connected to a 60 kV rural distribution network (with a short-circuit level of 745 MVA) via a 3-winding transformer to limit the fault level of the 15 kV busbars.

The STATCOM is a voltage source converter based reactive power compensator which may be used to generate controlled reactive power, and in this case the control system was arranged so that the wind farm generated at unity power factor but without the risk of self-excitation. The STATCOM (or ASVC) consisted of two 3-level GTO (gate turn-off thyristor) converters connected in a 12-pulse configuration to limit harmonics. A limited SHEM (selective harmonic elimination) switching pattern was also used to remove some of the lower-order harmonics although this, of course, led to increased magnitudes of some of the higher harmonics. The experimental installation worked well but it must be emphasised that, at the time of writing, the use of shunt connected power electronic compensators to provide reactive power compensation for induction generators is limited to one or two research/demonstration projects and is not common commercial practice.

4.3 Power electronic converters

Power electronic converters are currently used to interface some forms of renewable generation and energy storage devices to distribution networks, and their use is likely to increase in the future. The development of these converters is benefiting from the rapid advances in power semiconductor switching devices and in the progress being made in the design and control of variable-speed drives for large motors. One obvious application of a power electronic converter is to invert the DC generated from some energy sources (e.g. photovoltaics, fuel cells or battery storage) to 50/60 Hz AC. However, converters may also be used to decouple a rotating generator from the network and so potentially increase the efficiency of the operation of the prime mover by ensuring it operates at its most efficient speed for a range of input power. This is one of the arguments put forward to support the use of variable-speed wind turbines but is also now being proposed for some forms of small hydro-generation. Another advantage of variable-speed operation is the reduction in mechanical loads possible by making use of the flywheel effect of the variable-speed generating set to store energy during transients. However, large power electronic converters do have a number of disadvantages, including (i) significant capital cost and complexity, (ii) electrical losses (which may include a considerable element independent of output power) and (iii) the possibility of injection of harmonic currents into the network.

Figure 4.27 shows a converter system typically used to control a large variable-speed wind turbine. Operation is possible over a wide speed range as all the power is rectified to DC and flows through the converter circuit. Therefore, with typical losses of 2–3% in each converter it may be seen that, at full load, some 4–5% of the output power of the generator may be lost. These losses may be reduced significantly by passing only a fraction of the power through the electronic converter (Figure 4.28), but in these schemes the variations in speed are limited. In the doubly fed wound rotor induction machine system, shown in Figure 4.28, the converter controls the frequency of the rotor current and hence its speed. With a 4-quadrant converter, operation is possible at above and below synchronous speed. Various forms of these slip-recovery systems were used on the multi-megawatt prototype wind turbines constructed as part of national research programmes in the early 1980s. In general, these turbines did not perform well for a number of reasons not connected with the variable-speed converter equipment and the concept was not pursued. However, recently commercial 1.5 MW 60 m diameter wind turbines have been offered for sale with IGBT (insulated gate bipolar transistor), PWM (pulse width modulated) converters in the rotor circuits. A very narrow speed range is also possible using a wound rotor induction machine by

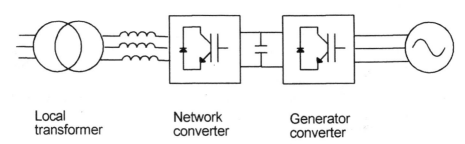

Local   Network   Generator
transformer   converter   converter

Figure 4.27   Wide-range variable-speed power electronic converter

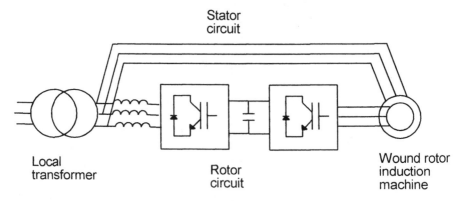

Stator
circuit

Local   Rotor   Wound rotor
transformer   circuit   induction
                                      machine

Figure 4.28   Narrow-range variable-speed power electronic converter

mounting a controllable resistor on the rotor but, in this case, the slip energy is not returned to the network but is lost as heat.

4.3.1 Voltage source converters

A very large number of concepts and power electronic circuits have been used or proposed for the connection of small generators to the network. These have taken advantage of the operating characteristics of the various semiconductor switching devices which are available and, as improved switches are developed, then the circuits will continue to evolve. Some years ago, naturally commutated, thyristor-based invertors were common and can still be found in service in some photovoltaic installations. This type of equipment has the advantage of low electrical losses but its power factor is determined by the voltage of the DC link and its harmonic performance is poor. Simple theory suggests that a six-pulse inverter operating at fundamental frequency will inject a 5th harmonic current of some 1/5th (20%) of output, a 7th harmonic current of some 1/7th (14%) of output and so on. As well as the possible expense of filters, a converter with a poor harmonic performance requires time-consuming and expensive harmonic studies to ensure that its impact on the network will be acceptable. Such inverters are now rarely used in new embedded generation schemes.

Most modern converters use some form of voltage source converter, which, as the name implies, synthesise a waveform from a voltage source. Figure 4.29 shows a typical voltage source network converter. Power from the energy source is supplied to the DC bus which, in the simplest arrangement, is held at a constant voltage by the operation of the energy source control system. The inverter then synthesises a voltage waveform which is used to inject current into the network across the coupling reactor (X_c) and the reactance of the local transformer. Thus, the basic operation of the voltage source converter is very similar to that of the synchronous generator (Figure 4.4 and eqns (4.3) and (4.4)). The real power injection (P) is controlled mainly by the phase angle of the

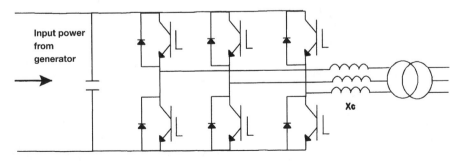

Figure 4.29   Six-pulse two-level IGBT voltage source converter

synthesised waveform with that of the network while the reactive power (Q) is controlled by the relative magnitudes of the two waveforms. If the DC bus voltage is fixed it may be seen that the inverter must control both the phase and magnitude of the synthesised waveform for independent control of real and reactive power flows. If current harmonics are not to be injected into the network then the inverter must synthesise a close to perfect sine wave.

The requirement to synthesise an approximation to a sine wave from a DC voltage source is found in many areas of power electronics, and the various modulation strategies are well known. They include the following:

- square wave inversion, which is the most basic approach, requiring the fewest switching operations and using a very simple control, but generating the maximum harmonics
- carrier modulated techniques, which compare a reference signal with a trigger signal. The most well known of these is sinusoidal pulse width modulation (PWM), which is easily implemented in hardware
- hysteresis control, in which the output is controlled to within reference bands either side of the desired wave
- programmed PWM (sometimes known as selective harmonic elimination, SHEM), in which sophisticated optimisation strategies are used, off-line, to determine the required switching angles to eliminate certain harmonics.

For the simple two-level circuit of Figure 4.29 a sinusoidal current can only be injected if the IGBTs are switched rapidly. This leads to electrical losses, which may not be as commercially significant in a large motor drive as they will be in an embedded generation scheme. Thus for large generators alternative arrangements may be considered including the use of multilevel inverters to combine a number of voltage sources or by combining multiple inverters together through transformers of differing vector groups to form multiphase inverters. The technique chosen will depend on an economic appraisal of the capital cost and cost of losses and, as technology in this area is developing rapidly so the most cost-effective technique at any rating will change over time. Future developments of power electronic converters for embedded generation plant are likely to involve the use of soft-switching converters to reduce losses; resonant converters are already used in small photovoltaic generators, and may include other converter topologies which eliminate the requirement for the explicit DC link.

The control schemes used on converters for embedded generators are similar to those used on large machine drives and are often based on space vector theory using some form of two-axis transformation [15]. These allow rapid and independent control of real and reactive power injected into the network.

IEC 909 requires that power electronic motor drives which can operate regeneratively are considered for their contribution to initial short-circuit current (both symmetrical and initial peak current) but not to any sustained short-circuit current. They are represented as an equivalent motor with a locked rotor current of 3× rated current. This would appear to be an appropriate, if rather conservative, assumption for the rating of switchgear in the presence of embedded generation plant. The CIGRE report on embedded generation [16] indicates that a fault current only equal to rated current may be expected and, of course, the actual behaviour of the inverter will depend both on its power circuit and control system. As with induction generators, the inability of power electronic converters to provide sustained fault current (unless this feature is specifically designed into the converter) for the operation of protective relays has important consequences for protection systems.

4.4 References

1 HINDMARSH, J.: 'Electrical machines and their applications' (Pergamon Press, Oxford, 1970)
2 MCPHERSON, G.: 'An introduction to electrical machines and transformers' (John Wiley and Sons, New York, 1981)
3 GRAINGER, J.J., and STEVENSON, W.D.: 'Elements of power systems analysis' (McGraw Hill, New York, 1982)
4 WEEDY, B., and CORY, B.J.: 'Electric power systems' (John Wiley and Sons, Chichester, 1998)
5 MOHAN, N., UNDELAND, T.M., and ROBBINS, W.P.: 'Power electronics, converters applications and design' (John Wiley and Sons, 1995)
6 HURLEY, J.D., BIZE, L.N., and MUMMERT, C.R.: 'The adverse effects of excitation system Var and power factor controllers'. Paper PE-387-EC-1–12–1997 presented at the IEEE Winter Power Meeting in Florida
7 IEEE COMMITTEE REPORT: 'Dynamic models for steam and hydro turbines in power system studies', *IEEE Transactions on Power Apparatus and Systems*, 1973, **92**, (6), pp. 1904–1915
8 IEC 909: 'Short circuit current calculation for three-phase a.c. systems'. International Electrotechnical Commission, 1988
9 ELECTRICITY ASSOCIATION: 'Procedure to meet the requirements of IEC 909 for the calculation of short circuit currents in three-phase AC power systems'. Engineering Recommendation G74, 1992
10 GRIFFITH SHAN, M.: 'Modern AC generator control systems, some plain and painless facts', *IEEE Transactions on Industry Applications*, 1976, **12**, (6), pp. 481–491
11 HEIER, S.: 'Grid integration of wind energy conversion systems' (John Wiley and Sons, Chichester, 1998)
12 ALLAN, C.L.C: 'Water-turbine driven induction generators', *Proceedings of the IEE*, Part A, Paper No. 3140S, December 1959, pp. 529–550
13 WAGNER, C.F. and EVANS, R.D.: 'Symmetrical components' (McGraw-Hill, New York, 1933)

14 SOBRINK, K.H., JENKINS, N., SCHETTLER, F.C.S., PEDERSEN, J., PEDERSEN, K.O.H., and BERGMANN, K.: 'Reactive power compensation of a 24 MW wind farm using a 12-pulse voltage source converter'. CIGRE International Conference on Large High Voltage Electric Systems, 30 August–5 September 1998
15 JONES, R. and SMITH, G.A.: 'High quality mains power from variable speed wind turbines'. IEE Conference on Renewable Energy – Clean Power 2001, 17 November 1993
16 CIGRE STUDY COMMITTEE No. 37: ' Impact of increasing contributions of dispersed generation on the power systems'. Final report of Working Group 37, 23 September 1998. To be published in *Electra*

Novar Wind Farm, Ross-shire, Scotland
The wind farm consists of 34 500 kW, single speed, stall regulated wind turbines.
It is connected into the local utility at 33 kV.

(Source: National Wind Power)

Chapter 5

Power quality

Over the last 10 years both operators and customers have taken an increasing interest in the quality of the power, or more precisely the quality of the voltage, which is delivered by distribution networks . This interest has been stimulated by a number of factors, including [1]:

- increasingly sensitive load equipment, which often includes computer-based controllers and power electronic converters which can be sensitive to variations in voltage magnitude, phase and frequency
- the proliferation of power electronic switching devices on the network, including the power supplies of small items of equipment such as PCs and domestic equipment, and the rise in background harmonic voltage levels on the electricity supply network
- increased sophistication of industrial and commercial customers of the network and concern over the effect, and commercial consequences, of perceived poor power quality on their equipment and hence production. This particularly concerns the effect of voltage dips or sags on continuous industrial processes.

Although the main issues of power quality are common to any distribution network, whether active or passive, the addition of embedded generation can have a significant effect and, as usual, increases the complexity of this aspect of distribution engineering. Some forms of embedded generation can introduce non-sinusoidal currents into the distribution network and so degrade power quality by causing harmonic voltage distortion while other types can cause unacceptable variations in RMS voltage magnitudes. Perhaps surprisingly, embedded generators may also improve network power quality by effectively increasing the short-circuit level of part of the distribution network.

Figure 5.1 shows a representation of how network power quality issues may be viewed. Figure 5.1*a* shows the various effects which may be considered to originate in the transmission and distribution networks and which can affect the voltage to which loads and generators are connected.

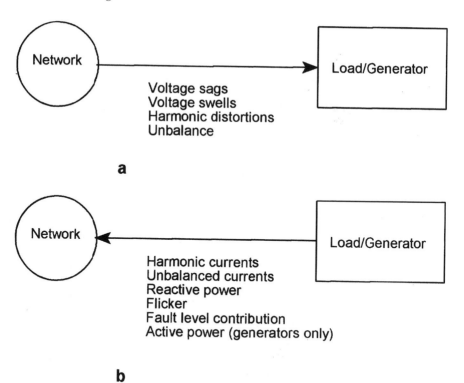

Figure 5.1   *Origin of power quality issues (after Dr. M. Weinhold)*

Thus, voltage sags (i.e. a decrease to between 0.1 and 0.9 per unit voltage for a period of up to 1 min), which are usually caused by faults on the transmission/distribution network, will impact on both loads and embedded generators. During severe voltage sags, spinning loads will tend to slow down and stall while rotating generating plant will accelerate, towards potential transient instability. In both cases the depressed voltage may cause contactors to open and voltage sensitive control circuits to mal-operate. Voltage swells are transient rises in power frequency voltage, usually caused by faults or switching operations, but are less common than sags. Ambient harmonic voltage distortion is increasing in many power systems, and in the UK it is not uncommon to find levels at some times of the day in excess of the level which is considered desirable by network planners [2]. Harmonic voltage distortions will have a similar effect of increasing losses in rotating machines whether they are operated as generators or motors and may also disturb the control systems of power electronic converters. It is common practice to use power factor correction capacitors with induction generators and these will, of course, have a low impedance to harmonic currents and the potential for harmonic resonances with the inductive reactance of other items of plant on the network. Network voltage unbalance will also affect rotating machines by increasing losses and introducing torque ripple. Voltage unbalance can also cause power converters to inject unexpected harmonic currents back into the network unless their design has included consideration of an unbalanced supply.

The response of embedded generators to disturbances originating on the transmission/distribution network will be broadly similar to that of large loads. However, large industrial loads will tend to be connected to strong networks with high short-circuit levels, and in general better power quality, while a distinguishing feature of embedded generation using some types of renewable energy is that they are connected to weak networks and the rating of the generation can be a significant fraction of the network short-circuit level.

Figure 5.1*b* gives a list of how loads or an embedded generator may introduce disturbances into the distribution network and so cause a reduction in power quality. Thus, either a load or a generator which uses a power electronic converter may inject harmonic currents into the network. Similarly, unbalanced operation of either a load or generator will lead to negative phase sequence currents being injected into the network which, in turn, will cause network voltage unbalance. As has been discussed earlier, generators can either produce or absorb reactive power while exporting active power and, depending on the details of the network, load and generation, this may lead to undesirable steady-state voltage variations. Voltage flicker refers to the effect of dynamic changes in voltage caused either by loads (arc-furnaces are a well known example) or by generators, e.g. wind turbines. Both spinning motor loads and embedded generators using directly connected rotating machines will increase the network fault level and so will affect power quality, often to improve it. Only generators are capable of supplying sustained active power which may lead to excessive voltages on the network. There is considerable similarity between the power quality issues of embedded generators and large loads and, in general, the same standards are applied to both.

The effect of a 'perfect' embedded generator on network power quality is illustrated in Figure 5.2. The 'perfect' generator is considered to supply perfectly sinusoidal current. Figure 5.2*a* shows simply a distorting load connected to a distribution system which is represented by a source impedance, $\overline{Z}_s(h)$, connected to an infinite busbar. The source impedance varies with frequency and so is written as a function of the harmonic number (h). The distorting load can be considered as a source of distorted current, $\overline{I}_d(h)$. Neglecting other loads, and using the principle of superposition, then the voltage distortion introduced by this load is simply found by applying Ohm's Law and finding the voltage distortion across the source impedance:

$$\overline{V}_d(h) = \overline{I}_d(h) \times \overline{Z}_s(h) \tag{5.1}$$

Figure 5.2*b* shows the 'perfect' embedded generator added to the network. If the principle of superposition is again used to investigate the effect of the distorted current then the voltage source of the embedded generator is considered as a short-circuit. Thus, the effect of adding the 'perfect' generator is to raise the short-circuit level of the network and effectively to introduce the generator impedance $\overline{Z}_g(h)$ in parallel with the source impedance $\overline{Z}_s(h)$. For the same injected current distortion $\overline{I}_d(h)$ the resultant voltage distortion is now $\overline{V}'_d(h)$ which, in this simple example, is merely

$$\overline{V}'_d(h) = \overline{V}_d(h) \times \overline{Z}_g(h)/(\overline{Z}_s(h) + \overline{Z}_g(h)) \tag{5.2}$$

The nature of the distorted load current determines which impedance is required. For fundamental current then, $\overline{Z}_s$ is merely the source 50/60 Hz impedance, while $\overline{Z}_g$ may need to include either the transient or synchronous reactance of the generator depending on the frequency of variations in $\overline{I}_d$. A similar representation can be used for calculations of harmonic voltages, in which case harmonic impedances are required. For calculation of unbalanced voltages, negative phase sequence impedances must be used.

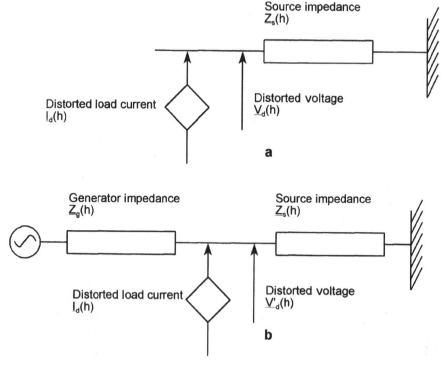

Figure 5.2   Effect of a 'perfect' embedded generator on network power quality

Of course, if the embedded generator itself produces distorted current (e.g. a small PV inverter using a line commutated bridge), then these will add to any distorted current injected by the load. This is shown in Figure 5.3, rather simply with the internal impedance of the load and generator ignored, to indicate that the voltage distortion $\overline{V}''_d$ is now increased to

$$\overline{V}''_d = \overline{Z}_s \times (\overline{I}_d + \overline{I}_g) \tag{5.3}$$

The addition of the distorted currents must take account of their frequencies and relative phases. If the phase relationship between harmonic currents is not known, then a root of the sum-of-squares addition may be appropriate.

Eqns. (5.1)–(5.3) indicate the main effects on power quality of adding distributed generation to a distribution network. In practice, more realistic models with more complex software programes are used. Harmonic analysis programs are well established to investigate how harmonic voltage distortion is affected by injection of harmonic currents [3], although there remains considerable difficulty with establishing harmonic impedances for items of plant and in particular in deciding on suitable representation of loads. Loads are an important source of damping of harmonic resonances and so can affect the magnitudes of network harmonic voltages considerably. Recently programs have begun to be available to predict the effect of embedded generation plant on voltage flicker, although so far these are specific to particular generation technologies [4]. If necessary, electro-magnetic simulations may be undertaken to examine the response of the network to embedded generation in the time domain. However, as mentioned in Chapter 3, these programs require skill for their effective use.

Wind turbines are a good example of embedded generation plant for which power quality considerations are important. Individual units can

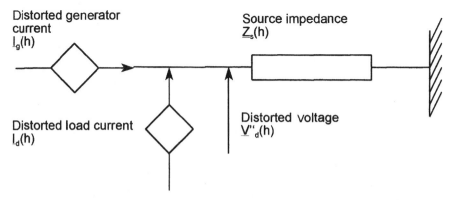

Figure 5.3   Effect of an embedded generator producing a distorted current on network power quality

be large, up to 1.5 MW, and are often connected to distribution circuits with a high source impedance. The turbines will either use induction generators (fixed rotor speed) or power electronic converters (variable rotor speed). For those designs which use power electronic converters the issues of harmonic distortion of the network voltage must be carefully considered while the connection of fixed-speed turbines to the network needs to be managed carefully if excessive transients are to be avoided. During normal operation wind turbines produce a continuously variable output power. The power variations are mainly caused by the effects of turbulence in the wind and tower shadow – the reduction in wind speed close to the tower. These effects lead to periodic power pulsations at the frequency at which the blades pass the tower (typically around 1 Hz for a large turbine), which are superimposed on the slower variations caused by meteorological changes in wind speed. There may also be higher-frequency power variations (at a few Hz) caused by the dynamics of the turbine. Variable-speed operation of the rotor has the advantage that many of the faster power variations are not transmitted to the network but are smoothed by the flywheel action of the rotor. However, fixed-speed operation, using a low-slip induction generator, will lead to cyclic variations in output power and hence network voltage.

Davidson [5] carried out a comprehensive, two year, measurement campaign to investigate the effect of a wind farm on the power quality of the 33 kV network to which it was connected. The 7.2 MW wind farm of 24 × 300 kW fixed-speed induction generator turbines was connected to a weak 33 kV overhead network with a short-circuit level of 78 MVA. Each wind turbine was capable of operating at two speeds by reconnection of the generator windings, and power factor correction capacitors were connected to each unit once the turbine started generating. The wind turbines were located along a ridge at an elevation of 400 m in complex upland terrain and so were subject to turbulent winds. Measurements to assess power quality were taken at the connection to the distribution network and at two wind turbines. The results are interesting as they indicate a complex relationship of a general improvement in power quality due to the connection of the generators, and hence the increase in short-circuit level, and a slight increase in harmonic voltages caused by resonances of the generator windings with the power factor correction capacitors. In detail the measurements indicated the following:

- The operation of the wind farm raised the mean of the 33 kV voltage slightly but reduced its standard deviation. This was expected as, in this case, the product of injected active power and network resistance was approximately equal to the product of the reactive power absorbed and the inductive reactance of the 33 kV circuit. The effect of the generators was to increase the short-circuit level and so 'stabilise' the network but with little effect on the steady-state voltage.

- The connection of increasing numbers of induction generators caused a dramatic reduction in negative phase sequence voltage from 1.5% with no generators connected to less than 0.4% with all generators operating. This was, of course, at the expense of significant negative phase sequence currents flowing in the generators and associated heating and losses.
- The wind farm slightly reduced the voltage flicker measured at the point of connection. This was a complex effect with the generators raising the short-circuit level but also introducing fluctuations in current.
- There was a slight increase in total harmonic voltage distortion with the wind farm in operation, mainly caused by a rise in the 5th and 7th harmonics when the low-speed, high-impedance winding of each generator was in service. This increase was probably associated with a parallel resonance of the high-impedance winding of the generators and the power factor correction capacitors.

These results are typical of the experience of connecting wind farms, with large numbers of relatively small induction generators, on to rural distribution circuits. In a number of studies some aspects of power quality have been shown to be improved by the effective increase in the short-circuit level. However, the effect of embedded generation plant on the distribution network will vary according to circumstances, and each project must be evaluated individually. In particular, large-single generators, as opposed to a group of smaller generators, will require careful attention.

5.1 Voltage flicker

Voltage flicker describes dynamic variations in the network voltage which may be caused either by embedded generators or by loads. The origin of the term is the effect of the voltage fluctuations on the brightness of incandescent lights and the subsequent annoyance to customers [6]. Human sensitivity to variations of light intensity is frequency-dependent, and Figure 5.4 indicates the magnitude of sinusoidal voltage changes which laboratory tests have shown are likely to be perceptible to observers [7]. It may be seen that the eye is most sensitive to voltage variations around 10 Hz. The various national and international standards for flicker on networks are based on curves of this type. Traditionally voltage flicker was of concern when the connection of large fluctuating loads (e.g. arc furnaces, rock crushing machinery, sawmills, etc.) was under consideration. However, it is of considerable significance for embedded generation, which: (i) often uses relatively large individual items of plant compared to load equipment; (ii) may start and stop frequently; (iii) may be subject to continuous variations in input power

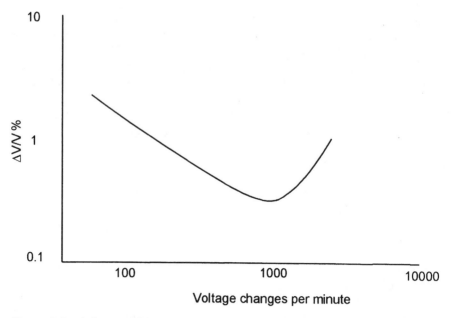

Figure 5.4   Influence of frequency and the perceptibility of sinusoidal voltage changes [7]

from a fluctuating energy source. Items of embedded generation plant which require assessment for potential nuisance caused by voltage flicker include:

- connection and disconnection of induction generators
- operation of wind turbines
- operation of photovoltaic generators.

Flicker is usually evaluated over a 10 min period to give a 'short-term severity value' P_{st}. The P_{st} value is obtained from a 10 min time series of measured network voltage using an algorithm based on the nuisance perceived by the human eye in fluctuating light [7]. P_{st} is linear with respect to the magnitudes of the voltage change but, of course, includes the frequency dependency indicated in Figure 5.4. Twelve P_{st} values may then be combined using a root of the sum of the cubes calculation to give a 'long-term severity value' P_{lt} over a 2 h period [6].

Some assessment of the effect of connection and disconnection of embedded induction generators can be made based on the established manual calculation procedure for large motors [6]. However, these manual techniques were not developed specifically for generators and there can be additional complications of deciding the power factor to assume for the distorted current being drawn through an operating anti-parallel thyristor soft-start unit and the effect of applied torque on the shaft of the generator. Therefore, for wind turbines an alternative procedure based on a series of measurements made at a test site has been proposed [8].

Determination of the voltage flicker caused by variations in real power output due to fluctuations in renewable energy sources is also difficult as this will depend on the resource, the characteristics of the generator and the impedance of the network. Simple measurement of voltage variations at the terminals of a test unit is not satisfactory as ambient levels of flicker in the network will influence the results, and the X/R ratio of the source impedance at the test site will obviously have a great impact on the outcome. For wind turbines, a procedure has been proposed where both voltage and current measurements are made of the output of a test turbine and used to synthesise the voltage variations which would be caused on distribution networks with defined short-circuit levels and X/R ratios of their source impedance. These voltage variations are then passed through a flicker algorithm to calculate the flicker which the test turbine would cause on the defined networks. When the installation of the particular turbine is considered at a point on the real distribution network these test results are then scaled to reflect the actual short-circuit level and interpolated for the X/R ratio of the point of connection. A correction is also applied for the annual mean wind speed [8].

If a number of generators are subject to uncorrelated variations in torque then their power outputs and effect on network flicker will reduce as

$$\frac{\Delta P}{P} = \frac{1}{\sqrt{n}} \times \frac{\Delta p}{p} \tag{5.4}$$

where n is the number of generators, P and p are the rated power of the wind farm and wind turbine, respectively, and ΔP and Δp are the magnitudes of their power fluctuation.

There is some evidence that on some sites wind turbines can fall into synchronised operation, and in this case the voltage variations become cumulative and so increase the flicker in a linear manner. The cause of this synchronous operation is not completely clear but it is thought to be due to interactions on the electrical system caused by variations in network voltage. However, experience on most UK sites, where the winds are rather turbulent, has been that any synchronised operation lasts only a short while before being disrupted by wind speed changes.

A range of permissible limits for flicker on distribution networks is given in national and international standards. Reference 6 specifies an absolute maximum value of P_{st} on a network, from all sources, to be 1.0 with a 2 h P_{lt} value of 0.6. However, extreme caution is advised if these limits are approached as the risk of complaints increases at between the 6th and 8th power of the change in voltage magnitude once the limits are reached, and the approximate method proposed in the same document is based on P_{st} not exceeding 0.5. Reference 9, which specifically excludes embedded generation from its scope, is significantly less stringent, specifying that over a one week period P_{lt} must be less than 1 for 95% of the time. Gardner [10] describes P_{st} limits from a number of utilities in the range 0.25–0.5 but also notes the rather different approach adopted by IEC 1000 [11], which uses a more complex methodology to allocate equitably the flicker capacity among all users of the network.

5.2 Harmonics

Embedded generation can influence the harmonic performance of distribution networks in a number of ways. Power electronic converters used to interface generation equipment can cause harmonic currents to flow, but conventional rotating plant (e.g. synchronous or induction generators) will alter the harmonic impedance of the network and hence its response to other harmonic sources. Further, the introduction of shunt capacitor banks used to compensate induction generators may lead to resonances.

Power electronic converters for interfacing large (> 1MW) embedded generation plant are still not widespread, with the possible exception of some designs of variable-speed wind turbines, but the future increased use of such equipment may be anticipated together with rapid changes in converter topologies and devices. In the last 5–10 years the voltage source converter has become common, but prior to that, line commutated, thyristor-based converters were used for some embedded generators. Figure 5.5 shows a simple 6-pulse, current source, line commutated bridge arrangement of the type used in some early photovoltaic and wind turbine schemes. This is a well established technology which has been widely used for industrial equipment and is the basis for contemporary high voltage direct current transmission plant. With a firing

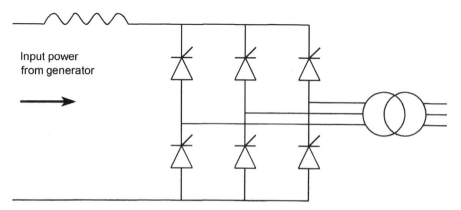

Input power
from generator

Figure 5.5   Line commutated converter

angle of the thyristors of between 0 and 90° the bridge acts as a rectifier with power flow from the network to the generation device, while inverter operation is achieved by delaying the firing angle of the thryistors to beyond 90°. The disadvantages of this technology are well known: (i) high characteristic harmonics and (ii) poor power factor. However, the arrangement is simple and robust with relatively low losses. It also has the advantage that islanded operation is not possible as the converter needs the commutating voltage of the network to operate.

Traditionally, state-owned electric power utilities have paid considerable attention to electrical losses as their investment appraisals were undertaken using long time horizons (20–40 years) and low discount rates (4–8%). In this financial environment electrical losses become very important as the costs incurred throughout the life of the plant are significant. Investment in industrial plant is assessed using much higher discount rates (perhaps 15–20%) with much shorter project lifetimes (perhaps 5–10 years) and so electrical losses are much less significant. Embedded generation falls somewhere between these two extremes, but currently schemes are often assessed using relatively short time-scales and high discount rates, reflecting the perceived risk in the projects and the source of funds. As embedded generation becomes a commercially well established activity then it is likely that investment decisions will be carried out in a manner similar to that used for other utility generating plant and so electrical losses will become more important. In that case, the use of line commutated converters, together with other low-loss converter designs, may become more attractive and the difficulties of high harmonic distortion accepted for the gains in efficiency.

The poor harmonic performance of the line commutated converter can be improved by using multiple bridges connected through transformers with two secondary windings with differing vector groups. This has the advantage of truly cancelling the low-order harmonics rather than merely shifting the harmonic energy higher up the spectrum, as occurs with the rapid switching of forced commutated converters. However, there is considerable additional complexity in the transformers and in the generation equipment. A possible requirement to filter some of the remaining harmonics to meet EMC regulations will remain. The power factor is effectively determined by the DC link voltage, which must be kept to a safe value to avoid commutation failure. Power factor correction capacitors and any filters are generally fitted to the high-voltage side of the transformer to avoid interaction with the converter.

The modern equivalent of Figure 5.5 was shown in Figure 4.29 [12]. This uses IGBTs switching at several kHz to synthesise a sine wave and so eliminate the lower-order harmonics. The harmonic performance quoted for such a converter is shown in Table 5.1 [12] and may be compared to that found in the previous generation of large industrial drives [13] which used line commutated converters. The reduction in low-order

Table 5.1   Typical harmonic currents from a sinuisoidal rectifier [12] compared to those of 6-pulse and 12-pulse large industrial AC drives [13]

Harmonic number	Network current harmonics (%)		
	Sinusoidal rectifier	6-pulse large industrial drive	12-pulse large industrial drive
1	100	100	100
3	1.9	–	–
5	2.8	21–26	2–4
7	0.5	7–11	1
11	0.16	8–9	8–9
13	0.3	5–7	5–7
17	0	4–5	0–1
19	0.125	3–5	0–1

harmonic currents is striking, although it may be seen that, in practice, the magnitude of the harmonic currents may differ significantly from that suggested by simple theory. In addition, the output current of the sinusoidal rectifier will have significant energy at around the switching frequency of the devices (i.e. in the 2–6 kHz region), and although these high-frequency currents are relatively easy to filter they do require consideration.

Thyristor soft-starts (Figure 4.22) are now commonly used to connect the induction generators used on fixed-speed wind turbines. When operating, these devices produce varying harmonics as the firing angle of the thyristors is altered to apply the network voltage to the generator. Generally the soft-start units are only used for a few seconds during the connection of the induction generator, and for this short period the effect of the harmonics is considered to be harmless and may be ignored [8]. If the antiparallel thyristors are not bypassed, but left in service, then their harmonic currents need to be assessed. This continuous use of the soft-start unit has been proposed in order to reduce the applied voltage and hence the losses in an induction generator at times of low generator output. However, as well as the complication of trying to deal with harmonics which change with the thyristor firing angle, there are also potential difficulties with the variation of the applied voltage to the generator altering the dynamic characteristics of the drive train.

The harmonic impedance of a distribution network will be altered by the addition of embedded generation. Arrilaga *et al.* [3] suggest that synchronous generators may be modelled as a series combination of resistance and reactance,

$$\overline{Z}_{gh} = Rh^{1/2} + jX_d''h \tag{5.5}$$

where R is derived from the generator power losses, X_d'' is the generator subtransient reactance and h is the harmonic number.

Transformers may be represented in the same way:

$$\overline{Z}_{th} = Rh^{1/2} + jX_t h \qquad (5.6)$$

where R is derived from the transformer losses and X_t is the transformer short-circuit reactance.

A rather complex representation of induction machines is proposed in Reference 3, but a simpler approximation is to use the negative phase sequence equivalent impedance and keep the X/R ratio of the impedance constant with respect to harmonic number.

A single embedded induction generator with its local power factor correction capacitor and transformer can be represented as shown in Figure 5.6 Using typical data for a 600 kW generator, the harmonic impedance looking into the transformer terminals is shown in Figure 5.7. Without any power factor correction the impedance of the generator and transformer is well behaved, increasing with harmonic number. The effect would be to lower the harmonic impedance of the distribution circuit to harmonics of low order. However, the introduction of the power factor capacitors has a considerable effect on the behaviour of

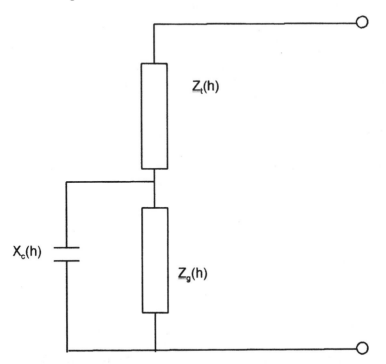

Figure 5.6   *Harmonic model of an induction generator with local transformer and power factor correction capacitors*

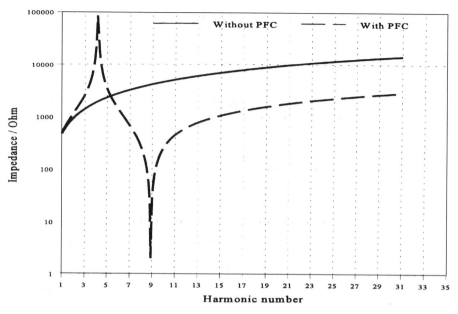

*Figure 5.7   Harmonic impedance of model of induction generator with local
transformer and power factor correction capacitors*

the circuit. It may be seen immediately that there is a series resonance between the transformer impedance and the capacitors (at around the 9th harmonic) and a parallel resonance between the generator and the capacitors (between the 4th and 5th harmonics). The series resonance may lead to excessive currents flowing in the capacitors, while the parallel resonance will increase the harmonic voltage distortion seen on the network. Thus any significant source of harmonics below the 11th in the distribution network (e.g. from other embedded generators, industrial loads or reactive power compensators) would be likely to interact with this network, and a comprehensive harmonic study may be required.

5.3 Voltage unbalance

Three-phase induction machines have a low negative phase sequence impedance (see Figure 4.24) and so will draw large currents if their terminal voltage is unbalanced. This leads to overheating and also ripple on the shaft torque. It is a reasonably common experience for small embedded induction generators to experience nuisance tripping on rural UK 11 kV distribution networks caused by network voltage unbalance. Unless special arrangements are made, small embedded generators may be designed for a network voltage unbalance (negative phase sequence voltage) of 1% with the unbalanced current protection set accordingly.

Synchronous machines may also be sensitive to network voltage unbalance as their damper windings will react in the same way as the squirrel cage of the induction generator. Power electronic converters are likely to respond to voltage unbalance by an increase in non-characteristic harmonics and possible nuisance tripping.

At present, the majority of embedded generators are three-phase and so do not cause an increase in voltage unbalance on the network. However, if domestic CHP or photovoltaic systems become common then there is an obvious problem that those houses with such equipment will load the distribution network less than those only with load. In the UK, only single-phase supplies are offered to most domestic dwellings and so if domestic generation becomes widespread then distribution utilities will be faced with the task of balancing the low-voltage feeders to ensure that each phase is approximately equally loaded and the neutral current is minimised for this new operating condition.

5.4 Summary

The main potential impacts of embedded generation on network power quality are voltage flicker and harmonic voltage distortion. If large numbers of single-phase generators are connected in the future then voltage unbalance may also become significant. The effect of embedded generation may be to improve network power quality by raising the short-circuit level, or cause it to deteriorate by introducing distorted current. The use of embedded generation also increases the amount of relatively large plant on the distribution network and so increases the complexity required in any modelling studies. The conventional techniques for assessing power quality on passive distribution networks are generally adequate for embedded generation, but there is a move to develop international standards and type testing procedures so that embedded generation plant can be assessed for its impact on any distribution network rapidly and with minimum cost [14]. Embedded generation may also suffer from existing poor distribution network power quality, and this aspect needs careful attention, especially for renewable energy schemes in rural areas. It has been found from experience that, at times in rural areas (before the connection of embedded generation), existing distribution network harmonic voltage distortion, voltage flicker and voltage unbalance can all exceed the desired values.

5.5 References

1　DUGAN, S., MCGRANAGHAN, M.F., and BEATY, H.W.: 'Electrical power systems quality' (McGraw Hill, New York, 1996)
2　ELECTRICITY ASSOCIATION: 'Limits for harmonics in the UK electricity supply system'. Engineering Recommendation G5/3, 1976
3　ARRILAGA, J., SMITH, B.C., WATSON, N.R., and WOOD, A.R.: 'Power system harmonic analysis' (John Wiley and Sons, Chichester 1997)
4　BOSSANYI, E., SAAD-SAOUD, Z., and JENKINS, N.: 'Prediction of flicker caused by wind turbines', *Wind Energy*, 1998 **1**, (1), pp. 35–50
5　DAVIDSON, M.: 'Wind farm power quality and network interaction'. Proceedings of the 18th British Wind Energy Association Conference, Exeter University, 25–27 September 1996 (Mechanical Engineering Publications), pp. 227–231
6　ELECTRICITY ASSOCIATION: 'Planning limits for voltage fluctuations caused by industrial, commercial and domestic equipment in the United Kingdom'. Engineering Recommendation P.28, 1989
7　MIRRA, C.: 'Connection of fluctuating loads'. International Union of Electroheat, Paris, 1988
8　INTERNATIONAL ELECTROTECHNICAL COMMISSION: 'Power quality requirements for grid connected wind turbines'. Report of Working Group 10 of Technical Committee 88, International Electrotechnical Commission, 1998
9　BRITISH STANDARD BS EN 50160: 'Voltage characteristics of electricity supplied by public distribution systems'. British Standards Institution, 1995
10　GARDNER, P.: 'Experience of wind farm electrical issues in Europe and further afield'. Proceedings of the 18th British Wind Energy Association Conference, Exeter University, 25–27 September 1996 (Mechanical Engineering Publications), pp. 59–64
11　INTERNATIONAL ELECTROTECHNICAL COMMISSION: 'Electromagnetic compatibility (EMC) Part 3: Limits – Section 7: Assessment of emission limits for fluctuating loads in MV and HV power systems – basic EMC publication'. IEC 1000–3–7, 1996
12　JONES, R., and SMITH, G.A.: 'High quality mains power from variable speed wind turbines'. IEE Conference on Renewable Energy – Clean Power 2001, 17 November 1993
13　FRANKENBURG, W., and GABRIEL, U.: 'High power converter-fed drives: treatment of supply system disturbances and power factor', Siemens Energy and Automation, Volume X, Special issue on *Large Electric Motor A.C. Variable Speed Drives*, October 1988, pp. 96–105
14　TANDE, J.O., and JENKINS, N.: 'International standards for power quality of dispersed generation'. Paper 4/7 CIRED International Conference on Electricity Distribution, Nice, France, 1–4 June 1999

Chapter 6

Protection of embedded generators

6.1 Introduction

Most protection systems for distribution networks assume power flows from the grid supply point to the downstream low-voltage network. This approach simplifies the problems associated with controlling the voltages at the loads and helps maintain the quality of supply. Protection is normally based on overcurrent relays with settings selected to ensure discrimination between upstream and downstream relays (see Figure 6.1). (Note that the figure incorporates the IEEE/ANSI designation numbers for a time-delayed phase overcurrent relay (51) and a time-delayed earth fault overcurrent relay (51N) [2, 5]) A fault on a downstream feeder must be cleared by the relay at the source end of the feeder. It must not result in the operation of any of the relays on an upstream feeder unless the downstream relay fails to clear the fault. If this occurs, relays on the immediately adjacent upstream feeders should operate and clear the fault. This will result in a blackout to a part of the network that should not have been affected by the fault [1, 3, 4].

If generation is embedded into the distribution system, the fault current seen by a relay may increase or decrease depending on its location and that of the fault and the generators (see Figure 6.2). To ensure appropriate co-ordination between relays, grading studies must take into account the maximum and minimum infeed from all the embedded generators and the grid supply point. This may create grading problems due to the intermittent nature of most of the embedded infeeds.

The main protection problem associated with the use of an embedded generator is not related to the protection of the generator but the protection of the intertie between the generator and the utility network. The relaying scheme used to protect this intertie must correctly co-ordinate with the network protection under all possible operating conditions. This requires knowledge about the topology of the utility network and the operating scenarios under which it may operate.

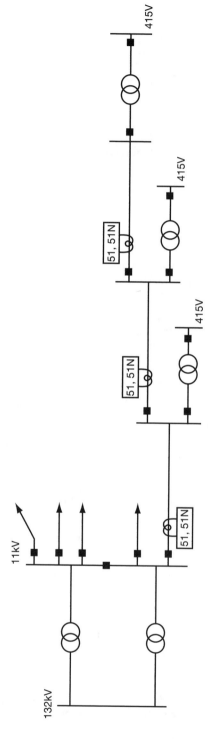

Figure 6.1   Overcurrent relays applied to radial distribution network

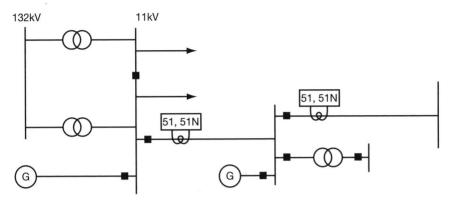

Figure 6.2   Overcurrent relays applied to radial distribution network with embedded generation

Most HV distribution networks operate as an open ring and are protected using overcurrent and earth fault relays. To reduce costs, small generators or industrial consumers with restricted levels of co-generation are often directly connected to a feeder, i.e. no switchgear is included at the tee point (see Figure 6.3). Multiple tapping of a feeder is generally accepted but in some cases the number of direct tee connections may be restricted.

Complex network configurations involving feeders with multiple directly connected generators may be difficult to protect using overcurrent relays and the fault clearance times may be excessive. In these situations, it may be necessary to connect small generators via switchgear with appropriate protection for the utility interface. Medium and larger embedded generators will always be connected to an 11kV or higher voltage busbar with appropriate switchgear and protection.

The power system of an industrial or commercial complex may include one or more generators to supply all or part of the load. Alternatively, the generators may be used to provide emergency power in the event of a failure of the normal utility supply. The former is referred to as an embedded generator and normally operates in parallel with the utility supply. The latter is described as an isolated system and is only allowed to operate when the industrial or commercial power system has been disconnected from the utility supply [2, 5, 6].

6.2 Protection schemes for isolated and embedded generators [2, 5, 6]

6.2.1 Single generator on an isolated network (50kVA–5MVA)

Many industrial or commercial power systems use a single isolated generator to supply emergency power if the utility supply fails. These generators are normally shut down and would be expected to operate

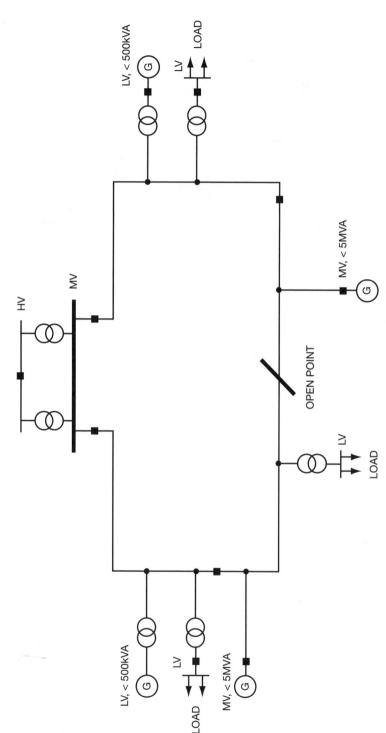

Figure 6.3   Open ring distribution network with LV and MV direct connected generators

infrequently and for short periods of time. The generator is only connected to the load when all other sources of supply have been disconnected, i.e. parallel operation is not allowed. Most generators of this type will start up automatically when the utility supply is lost and shut down when the supply is restored. A diesel engine or gas turbine would normally be used as the prime mover, and the rating of the generator would typically vary between 50kVA and 5MVA. Small machines (<500kVA) normally operate at low voltage and larger machines (500kVA–5MVA) at medium voltage. The definitions used in this chapter for low voltage (LV), medium voltage (MV) and high voltage (HV) are less than 1kV, between 1kV and 20kV, and greater than 20kV, respectively. In the UK, LV refers to 415V, MV to 3.3kV, 6.6kV and 11kV, and HV to 33kV, 66kV and 132kV.

A small generator (50–500kVA) connected to an isolated LV power system should be protected by a three-phase time-delayed overcurrent relay (51). If voltage signals are available at the line terminals, the over-current relay should be voltage controlled or restrained (IEEE/ANSI designation 51V). If the generator is wye connected and the neutral solidly earthed, the phase overcurrent relay should be energised using current transformers (CTs) installed in the phase conductors at the neutral terminals of the winding. An earth fault time-delayed overcurrent relay (51N) connected to a CT in the neutral should also be included in the protection scheme. The configuration of the protection scheme, assuming the windings are earthed wye, is shown in Figure 6.4*a*. If the generator is delta connected, the phase overcurrent relay should be connected to CTs on the line side of the winding terminals as shown in Figure 6.4*b*.

A synchronous generator (500kVA – 5MVA) connected to an isolated MV power network, with no other sources of generation, should be protected by a three-phase time-delayed voltage controlled overcurrent relay (51V), an earth fault time-delayed overcurrent relay (51N), a phase

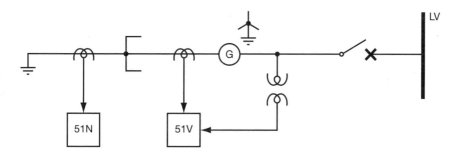

51V – 3 phase voltage controlled time–delayed overcurrent
51N – earth fault time–delayed overcurrent

Figure 6.4a   Protection for generator connected to isolated LV network. Generator: solidly earthed wye

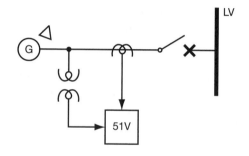

51V – 3 phase voltage controlled
time–delayed overcurrent

Figure 6.4b   *Protection for generator connected to isolated LV network.*
Generator: delta configuration

differential protection scheme for the stator winding (IEEE/ANSI designation 87), a loss of excitation relay (IEEE/ANSI designation 40), an under/overfrequency relay and an under/overvoltage relay. The configuration of the protection scheme is shown in Figure 6.5. The configuration assumes the windings of the generator are connected in a delta, and an earthing transformer is used to earth the isolated power network.

6.2.2   Generator operating in parallel with other generators on an isolated network

Industrial systems, or remote networks that require an uninterruptable supply, may incorporate multiple generators designed to operate in parallel with each other, but not in parallel with the utility supply. The size of each generator depends on the system demand but typically varies between 500kVA and 5MVA. Normally the prime mover is a diesel engine or gas turbine and the generator is connected to an MV network (MV = 1kV–20kV).

A generator (500kVA–5MVA) connected to an isolated power system that includes at least one other generator should be protected by a three-phase time-delayed voltage controlled overcurrent relay (51V), an earth fault time-delayed overcurrent relay (51N), a reverse power relay (32), a phase differential protection scheme for the stator winding (87), a loss of excitation relay (40), an under/overfrequency relay and an under/overvoltage relay. The loss of excitation relay is not required if the excitation system in the synchronous generator is unable to sustain short-circuit current or the generator is an asynchronous machine (induction generator). A typical protection scheme for this type of generator is

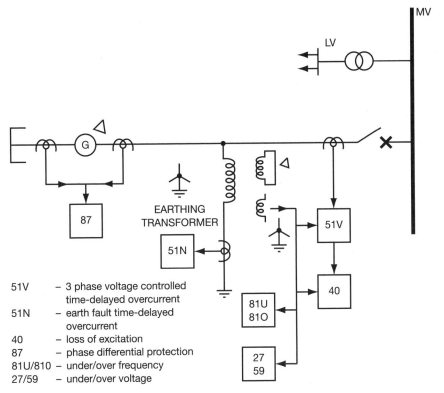

Figure 6.5 *Protection for synchronous generator connected to isolated MV network*

51V — 3 phase voltage controlled time-delayed overcurrent
51N — earth fault time-delayed overcurrent
40 — loss of excitation
87 — phase differential protection
81U/81O — under/over frequency
27/59 — under/over voltage

shown in Figure 6.6. The configuration assumes the windings are earthed wye.

6.2.3 *Generator embedded into utility network*

The protection requirements for a small generator (50–500kVA) operating in parallel with the LV utility network (LV < 1000V) is three-phase time-delayed voltage controlled overcurrent (51V), earth fault time-delayed overcurrent (51N), reverse power (32), under/overfrequency, under/overvoltage and in the UK loss of grid protection [5, 7].

The protection requirements for a generator (500kVA–5MVA) operating in parallel with the MV utility network is three-phase time-delayed voltage controlled overcurrent (51V), earth fault time-delayed overcurrent (51N), phase differential protection (87), reverse power (32), under/overfrequency, under/overvoltage and in the UK loss of grid protection. Synchronous machines also require loss of excitation protection (40). The configuration of the protection scheme is shown in Figure 6.7.

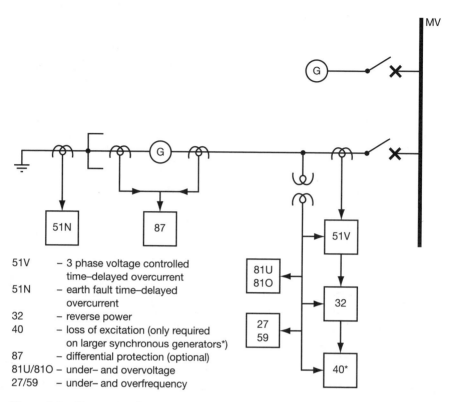

Figure 6.6   Protection for generator operating in parallel with other generators on an isolated MV network

The protection requirements for a generator (5–50MVA) operating in parallel with an HV utility network is three-phase time-delayed voltage controlled overcurrent (51V), earth fault time-delayed overcurrent (51N), phase differential protection (87), separate earth fault differential protection (87N), reverse power (32), under/overfrequency (81U, 81O), under/overvoltage (27,59), negative phase sequence overcurrent (46), unbalanced loading protection and in the UK loss of grid protection. Synchronous machines also require loss of excitation protection (40) and rotor protection (64). The configuration of the protection scheme is shown in Figure 6.8.

6.2.4 Protection requirements

The statutory obligation on an electricity utility is to protect the distribution network and the supplies to customers from excess current and earth leakage current. This requires appropriate protection between the sources of power and the points of supply. The settings applied to the

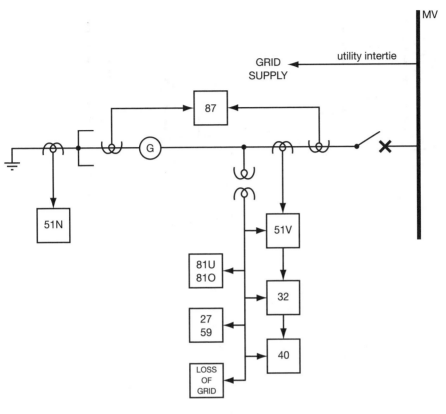

51V	– 3 phase voltage–controlled time–delayed overcurrent
51N	– earth fault time–delayed overcurrent
32	– reverse power
40	– loss of excitation
87	– differential
81U/81O	– under– and overfrequency
27/59	– under– and overvoltage
LOSS OF GRID	– requirement in UK

Figure 6.7   Protection for generator embedded into utility MV network

protection must take into account the network loads and the range of fault currents during both normal and abnormal operating conditions. An embedded generator is a source of power and consequently the utility has to ensure that adequate protection is applied to the generator–utility interface to safeguard the utility network and other customers. In the United Kingdom, the minimum protection requirements are specified in a document referred to as G59 [10]. This provides recommendations for the connection of individual generating plant to the UK utility distribution system and allows the utility to pass the minimum protection obligation to the generator without the need for a detailed technical study. The

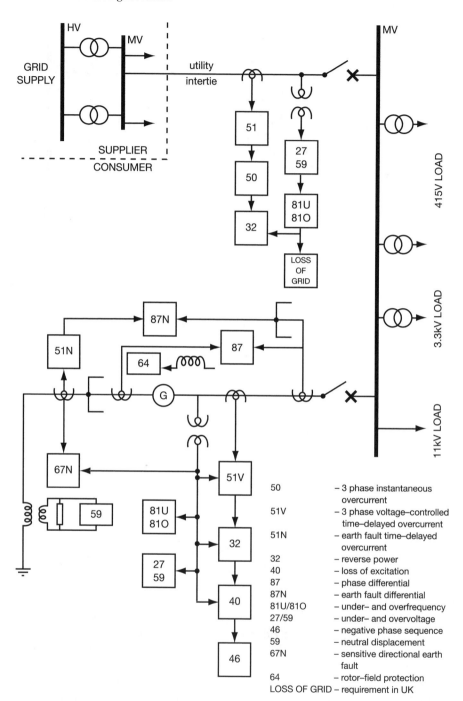

Figure 6.8   Protection for generator supplying industrial load and embedded into utility HV network

document defines the minimum requirements for safe operation of embedded plant with a rating of less than 5MW and for connection at voltages less than 20kV.

In addition to the protection associated with the generating plant, the public utility requires that the embedded generator provides adequate protection to satisfy all the following conditions:

(i)   The generator is prevented from connecting to the utility network unless all phases of the utility supply are energised and are operating within agreed settings.

(ii) The generator is disconnected from the utility network if a system abnormality occurs that results in an unacceptable deviation in voltage or frequency at the point of supply.

(iii) The generator is disconnected from the utility network if one or more phases of the utility supply is lost at the point of connection.

(iv) The generator must either be automatically or manually disconnected from the utility network if a failure of any supply to the protection equipment inhibits its correct operation.

The exact protection arrangements and the settings associated with the relays depend on the type and size of the generator installation and the requirements of the utility network. To satisfy the conditions listed above, the protection must include the detection of overvoltage, under-voltage, overfrequency, underfrequency and loss of grid. In addition most utilities recommend overcurrent, earth fault, reverse power and neutral voltage displacement.

G59 specifies that the minimum protection necessary to connect a small asynchronous generator (<150kVA) to an LV utility network is an under- and overvoltage relay on each phase and a single under- and overfrequency relay. The settings applied to these relays must ensure adequate protection of the utility network. However, it is not necessary for these relays to protect the generator, since this uses additional relays as described in Section 6.2.3. For generators with a capacity of greater than 150kVA, G59 specifies under- and overvoltage, under- and over-frequency and loss of mains protection. It also recommends, particularly for larger machines, neutral voltage displacement, overcurrent, earth fault and reverse power.

In the UK, all MV connections to a utility network must include overcurrent and earth fault protection at the point of metering. Add-itional protection is also required to provide time-delayed back-up and to resolve any problems resulting from the disconnection of the earth on the utility network or weak infeed from the generator. Possible options include the use of neutral voltage displacement protection, directional overcurrent relays, reverse power protection, and underimpedance relays.

6.3 Overcurrent protection

6.3.1 Overcurrent protection of the generator intertie [2,4,5]

The main function of the protection on the intertie between the generator and the utility network is to disconnect the generator if a fault on the network has not been cleared within an acceptable time. The relay protects the distribution system against excessive damage and prevents the generator from exceeding its thermal limits. If the generator is connected to a distribution system protected by overcurrent relays, the protection on the intertie is based on instantaneous and time-delayed overcurrent elements. The instantaneous elements are designed to only operate on close-up solid faults; they must remain stable for faults located anywhere else on the utility network. Inverse definite minimum time (IDMT) elements are designed to operate in a time that co-ordinates correctly with the other IDMT relays on the network [1]. If the current and time settings are too low, the relay may trip the generator unnecessarily on overloads, power swings and severe voltage disturbances, or trip the generator during a remote fault which should have been cleared by relays nearer the fault. If the current settings are too high, the relay may not operate on a fault that is within its zone of protection. This problem is exacerbated by the decaying characteristics of the generator fault current. To ensure correct operation, grading studies must be performed to confirm that the overcurrent protection associated with the generator intertie will co-ordinate correctly with all the relays on the network under all operating conditions. These studies must take into account worst-case conditions for all the protection relays on the network, i.e. a delayed trip with a failure of the circuit breaker, maximum and minimum infeeds, possible changes in the operating topology of the network and worst-case overload conditions for which the protection must remain stable. In some cases, the result of these studies may indicate that suitable grading margins are difficult or impossible to achieve and the user or system designer may decide that inverse or definite time overcurrent relays are inappropriate. A possible alternative is a voltage-dependent overcurrent relay, but this requires access to a voltage transformer.

Inverse time-delayed overcurrent (51) protection is normally used to detect phase and solid earth faults on the intertie. The operating time depends on the magnitude of the fault current, the pick-up level (current or plug setting), the time multiplier setting (TMS) and the chosen curve. The pick-up level is often referred to as the current, tap or plug setting or a minimum operating threshold. The chosen curve is selected from a set of curves specified in the IEC255 standard. The operating time equations (in seconds) for the:

- IEC255 standard inverse curve is $t = 0.14 \times \text{TMS} / (I^{0.02} - 1)$

- IEC255 very inverse curve is $t = 13.5 \times \text{TMS} / (I^{1.0} - 1)$
- IEC255 extremely inverse curve is $t = 80.0 \times \text{TMS} / (I^{2.0} - 1)$

where I is the current expressed as a multiple of the pick-up level.

Voltage dependent time-delayed overcurrent (51V) protection can be used to detect phase and solid earth faults on the intertie. The operating characteristic depends on the magnitude of the voltage; as the voltage decreases, both the pick-up level and the operating time are reduced. The exact behaviour depends on whether the relay is classified as voltage restrained or voltage controlled. In a voltage restrained relay the minimum current required to operate the relay at a given tap or operating setting drops in accordance with the reduction in voltage. In a voltage controlled relay the operating curve changes from a load to a fault characteristic when the voltage drops below a preset level, normally set between 60% and 80% of the rated voltage.

On a solidly grounded system, a phase-phase fault causes the phase-neutral voltage to reduce to approximately 50% of its nominal value. For a phase-earth fault the voltage reduction depends on the location of the fault; for close-up faults the reduction is large, but for remote faults it is small. This reduction in voltage can be used to control sensitive over-current elements, that would otherwise mal-operate on load.

Instantaneous phase overcurrent (50) protection provides immediate tripping of the generator when the current in the intertie exceeds a high set overcurrent threshold. The threshold must be set so that it can only detect close-up faults; it must not operate for a fault that should be cleared by a relay on the utility network. In some situations it may be preferable to use instantaneous voltage controlled phase overcurrent protection (50V); this allows instantaneous tripping when the phase current exceeds a threshold and the voltage is less than a preset value (60–80% of nominal voltage). If the intertie has been operating normally and the current suddenly increases, then a fault or a load change has occurred. If the operating threshold has been set higher than the maximum load current, then the disturbance is due to a fault. An undervoltage level detector helps decide whether the fault is close-up or remote. The voltage collapse on a close-up fault is normally significantly greater than the collapse due to a remote fault. Consequently, undervoltage control allows the instantaneous overcurrent elements to be set more sensitively than can be achieved without voltage control.

6.3.2 Example of how overcurrent protection can be applied to an LV connected generator [8]

An example of a grading study applied to an LV connected 800kW generator is described in Figure 6.9. The overcurrent protection at A on the 415V feeder is a three-phase standard inverse relay with high set

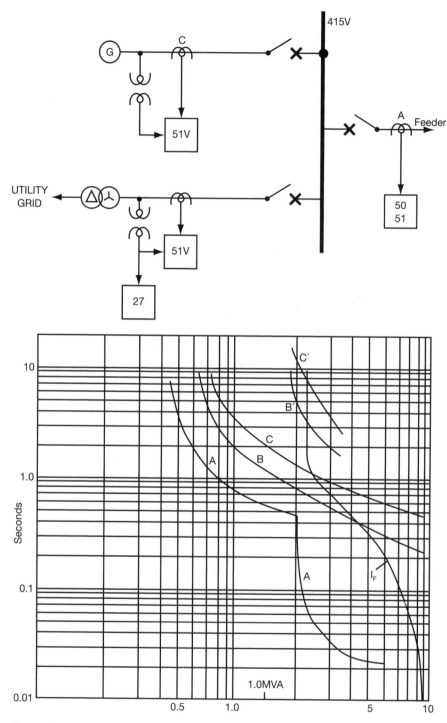

Figure 6.9   Grading study on an LV connected embedded generator

instantaneous elements. The inverse elements have a plug or operating setting of 400A and a time multiplier of 0.1. The instantaneous elements operate at 2000A. The overcurrent protection at B on the LV side of the 11kV/415V utility transformer is a three-phase voltage controlled standard inverse relay. The time multiplier is 0.125 and the plug setting is 600A when the voltage is less than 75% of nominal and 1725A when the voltage is greater than 75%. The overcurrent protection at C, on the generator terminals, is also a voltage controlled standard inverse relay. The time multiplier is 0.225 and the plug setting is 600A at V<75% and 1725A at V>75%.

For a phase fault on the 11kV intertie, the voltages on the LV side of the delta:wye transformer decrease below the 75% voltage setting associated with the relays at B and C. Consequently and with reference to the grading diagram, relay B operates and trips the intertie circuit breaker before any damage can occur to the transformer, i.e. the voltage controlled overcurrent relays provide adequate protection for phase faults on the 11kV intertie. The thermal withstand characteristic of the transformer is shown as the I_F curve in the grading diagram. For phase-to-earth faults on the 11kV intertie, none of the LV voltage signals decrease to below 90% and consequently the relays at B and C are now restricted to a 900A inverse overcurrent characteristic. The fault clearance time is unable to prevent damage to the transformer.

Undervoltage relays with a setting of 90% of nominal voltage and an operating time of 500ms are normally used in conjunction with the voltage dependent overcurrent relays to protect the generator intertie. However, for a phase-to-earth fault on this 11kV intertie, the voltages remain above the operating setting. Neutral voltage displacement protection, as shown in Figure 6.11, can detect the imbalance resulting from a phase-to-earth fault on the 11kV intertie. The main disadvantage of neutral voltage displacement is the requirement for wye connected voltage transformers on the 11kV side of the power transformer.

6.3.3 Negative sequence overcurrent

Negative sequence overcurrent elements are designed to detect phase-phase faults [1]. They can be set to be more sensitive than the phase elements because the negative sequence component of the current flowing into a balanced load is zero. Consequently, they only operate on the change in current caused by a phase-phase or phase-earth fault. They cannot detect the current flowing into a balanced load or a balanced three-phase fault. Earth fault overcurrent elements are normally set to be more sensitive than the negative sequence elements and would be expected to clear phase-to-earth faults.

6.3.4 Directional control of overcurrent elements

When fault current can flow in both directions, it may be necessary to supervise the overcurrent elements using directional elements. The overcurrent elements should be prevented from operating if the fault is in the reverse direction, i.e. behind the relay. The directional elements measure the direction of power flow by comparing the phase angle of the fault current with a reference obtained from the voltage signals.

6.4 Earth fault overcurrent protection

6.4.1 Methods of earthing the generator

The type of earth fault protection applied to the generator and its intertie depends on the method used to earth the generator and the utility network [1, 3, 5]. If a generator is connected to a utility network via a transformer with an earthed wye on the utility side, the outputs from the line current transformers on the intertie can be residually connected to earth fault overcurrent relays (see Figure 6.10). For an earth fault on the intertie or utility network, a part of the zero-sequence or residual fault current would flow around a loop which included the earthed wye winding of the generator transformer. An instantaneous earth fault relay (50N) could then be used to provide high-speed protection for close-up earth faults and a time-delayed relay (51N) to provide delayed earth fault protection for the intertie and back-up earth fault protection for the utility network. This method of earthing would normally be used when connecting a generator to a transmission or subtransmission network, but would not normally be used on a distribution network, where the utility side of the generator transformer would be delta connected. Most distribution circuits are operated with the neutral earthed at a single point, and this is normally at the grid supply point. Consequently, zero sequence current can only flow from this source to an earth fault; it cannot flow from the embedded generator to the fault.

If an earth fault occurs on a distribution feeder that includes an embedded generator operating via an unearthed transformer, the feeder will remain energised due to the infeed from the generator. During this situation and assuming all other sources of infeed have been disconnected the phase to earth voltage on the non-faulted phases may increase to be as high as the nominal phase-phase voltage. A zero sequence voltage relay (59N), connected to the utility side of the embedded generator transformer, will detect the neutral voltage displacement caused by an earth fault and immediately trip the embedded generator off the network. Neutral voltage displacement protection, as described in Figure 6.11, prevents damage due to the overvoltages and overcurrents resulting from a sustained earth fault.

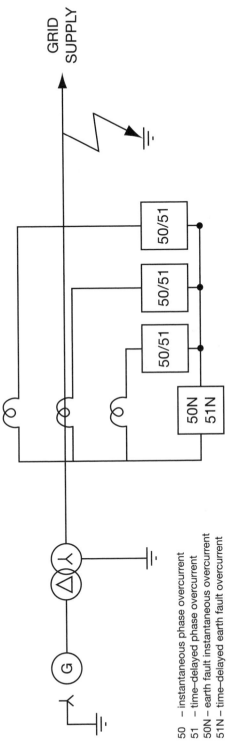

50   – instantaneous phase overcurrent
51   – time–delayed phase overcurrent
50N – earth fault instantaneous overcurrent
51N – time–delayed earth fault overcurrent

Figure 6.10   Earth and phase fault protection of intertie

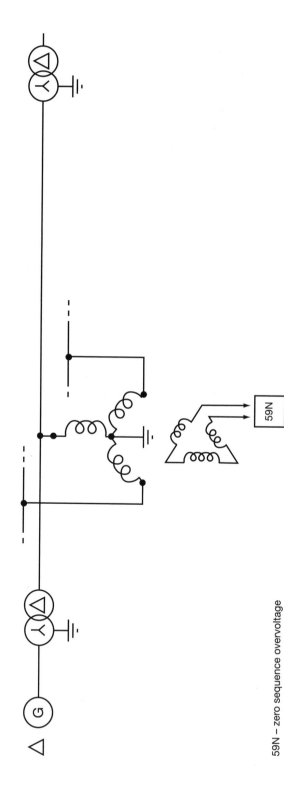

59N – zero sequence overvoltage

Figure 6.11   Neutral voltage displacement protection applied to intertie

6.4.2 Time-delayed earth fault overcurrent

For a generator connected to the utility network via a step-up transformer with an earthed wye on the utility side, a time-delayed earth fault overcurrent relay (51N) is normally included with the protection used to trip the circuit breaker on the generator intertie [1, 5]. The relay provides delayed clearance of earth faults on the intertie and back-up protection for earth faults on the utility network. For a directly connected generator, no step-up transformer, the relay also provides earth fault protection for the stator winding of the generator, but the operating time is slow due to the need to co-ordinate with earth fault relays on the utility network.

On small isolated systems operating with a directly connected generator, a time-delayed earth fault overcurrent relay (51N) used to monitor the current in the generator neutral, provides earth fault protection for the generator. The magnitude of the earth fault current depends on the method used to earth the generator and the system to which it is connected. Direct connection of the neutral of the generator to earth is normally restricted to low-voltage systems; connection through a low or high resistance is used at voltage levels above this value. With low-resistance earthing, the resistance is selected to provide an operating current of approximately rated current for a bolted phase-to-earth fault at the line terminals. The pick-up settings for the earth fault relay in the neutral should ideally be approximately one-fifth of this value of operating current and be at least equal to and preferably greater than the pick-up settings for the earth fault relays on the utility network.

6.4.3 Earthing of transformer connected generators

If the generator is connected to the utility network via a transformer, the stator winding of the generator and the associated transformer winding can be treated as an isolated system that is not affected by the earthing on the utility system; this assumes the transformer is not connected earthed wye : earthed wye. Due to the risk of overvoltages, it is not recommended that the generator system is left floating. A common practice in larger machines is to earth the neutral of the generator through a single-phase distribution transformer with a resistive load on the secondary. The resistor is designed to prevent transient overvoltages resulting from the effect of an arcing earth fault on the capacitance of the winding. The resistor is designed to discharge any bound charge in the circuit capacitance and its value, referred to the primary of the earthing transformer, should not exceed the impedance at power frequency of the total summated capacitance of all three phases of the generator, its connections and the generator transformer. Expressed differently, the resistor should be selected such that the current flowing into a solid phase-to-earth fault at the terminals of the generator should not be less than the residual capacitance current. In some countries fault currents of up to five times the above value are permitted [1, 3].

Example: Select an appropriate resistor for loading the secondary of an earthing transformer. Assume the generator is rated at 100MVA, 13.8kV, the generator transformer is rated at 13.8kV/144kV and the capacitance per phase of the generator winding, the generator transformer and the connecting leads are 0.2μF, 0.01μF and 0.005μF, respectively. The total residual capacitance is 0.645μF and its impedance at 50Hz is 5kΩ. If the effective impedance of the earthing resistor is made equal to the total residual capacitive impedance of 5kΩ, then with a generator terminal earth fault, the neutral current is $13\,800 \div (\sqrt{3} \times 5000) = 1.6$A. The actual fault current contains almost equal resistive and capacitive components and will be approximately 2.2A [1].

An earthing transformer with a primary knee point voltage of 1.3×13.8kV = 18kV and a turns ratio of 18000:250 is used for earthing the generator. The required equivalent earthing resistance is 5kΩ, and referred to the secondary the earthing resistance is 0.96Ω. If we assume that the winding resistance referred to the secondary is 0.16Ω, the required ohmic value of the loading resistor is 0.8Ω. The secondary current in the earthing transformer for a generator terminal fault is approximately 160A ($= 2.2$A $\times 18\,000 \div 250$). A current transformer of ratio 100:1 connected to a time-delayed overcurrent relay could be used to monitor the current in the secondary of the earthing transformer. The pick-up setting of the relay will be typically 5% of nominal current, i.e. 5% of 100A = 5A in this example. With this setting the relay should be capable of providing earth fault protection for approximately 90–95% of the winding. The relay is unable to protect 100% of the winding because for a close-up fault the generated emf that drives the current around the earth loop is very small, and consequently the earth fault current is small.

For example, if an earth fault occurs at a point 7% along the winding from the neutral, and if one ignores the capacitive charging current, the earth fault current in the secondary of the earthing transformer is $0.07 \times 18\,000 \times 13\,800 \div (\sqrt{3} \times 5000 \times 250) = 8$A. Consequently, the relay current is $1.6 \times$ pick-up setting and if the relay has an IEC standard inverse characteristic then the operating time is approximately 15 s × TMS. If the fault occurred at the terminals of the generator and the capacitive charging current was equal to the resistive current, the current in the secondary of the earthing transformer would be 160A. Consequently the relay current is $32 \times$ pick-up setting and the operating time is 1.95 s × TMS. Note: a TMS setting of about 0.3 should be adequate.

As an alternative to an overcurrent relay, a time-delayed voltage relay could be connected across the secondary winding of the earthing transformer. However, since the earthing system is designed to restrict the fault current to a low value, the fault current may not be significantly greater than the third harmonic current and consequently the voltage relay must be insensitive to the effects of third harmonic distortion. A voltage relay tuned to operate at the power frequency or designed to attenuate non-power frequency components must be used. With reference to the example, the ohmic value of the loading resistor is 0.8Ω, the current in the secondary of the earthing transformer for a terminal earth fault is approximately 160A and consequently the voltage across the relay is 128V.

For an earth fault at a point 7% along the stator winding from the neutral the current in the secondary is 8A and the voltage across the relay is 6.4V. For this example and with a setting of 6.4V the voltage relay would provide earth fault protection for 93% of the winding. Typical setting values are between 5V and 20V [1].

The voltage relay is time delayed to prevent mal-operation due to transient surges that arise on the utility network. These surges may pass through the interwinding capacitance of the generator transformer and into the neutral connection of the generator. In addition, high-speed operation is not required since the fault current is restricted to a low value by the grounding resistor. An operating time of 1 to 2 s at 10 times the voltage setting is normally appropriate.

6.4.4 Earthing of directly connected generators

Generators directly connected to a distribution system, no step-up transformer, are normally earthed though a resistance which will pass approximately rated current to a terminal fault. A current transformer mounted on the neutral/earth connector can be used to energise an instantaneous overcurrent relay with a setting of 10% of the maximum earth fault current. This is normally considered a minimum setting, since lower settings may result in mal-operation due to imbalance or transient surge currents arising from the distribution system. In some situations, a time-delayed overcurrent relay with a setting as low as 5% of the maximum earth fault current would also be connected to the neutral CT.

6.5 Differential protection of the stator winding

6.5.1 Operating principle

The best method of protecting the stator winding of a generator is to use a Merz–Price circulating current differential protection system [1, 11, 9]. The principle of operation can be explained with reference to Figure 6.12a.

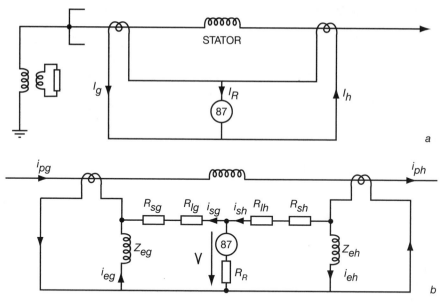

Figure 6.12   *Circulating current differential protection system (a) and equivalent circuit (b)*

If the current transformers (CTs) are ideal the functioning of the system is simple and can be described as follows:

for an external fault or a healthy winding, $\qquad I_g = I_h \qquad \therefore I_R = 0$
for an internal fault, $\qquad\qquad\qquad\qquad\quad I_g \neq I_h \qquad \therefore I_R \neq 0$

In practice, CTs have losses and different magnetic characteristics which may cause deviation from the ideal. Using the equivalent circuit of a CT and taking into account the secondary wiring impedance the circulating current system can be represented as the equivalent circuit shown in Figure 6.12b, where R_{sg} and R_{sh} are the winding resistances of CT$_g$ and CT$_h$, R_{lg} and R_{lh} are the lead resistances between CT$_g$ and the relay and CT$_h$ and the relay, and R_R is the burden impedance of the relay which is assumed to be purely resistive.

If an asymmetrical current signal is applied to a CT, the component of the signal that is normally described as a transient DC offset may cause saturation of the CT core. Saturation results in a reduction in the magnetising impedance and an increase in the exciting current. If the excitation characteristics of the balancing CTs in a differential scheme differ, or the CTs are unequally burdened, then the magnetic flux and the resulting excitation currents will not be equal. Consequently, even if the primary currents are equal, as is the case with an external fault or under normal operating conditions, the imbalance in the excitation currents will result in a spill that flows through the relay. To determine the maximum possible amplitude of the spill current assume one CT is fully saturated, i.e. its magnetising impedance is zero ($Z_{eh} = 0$) and the other CT is non-saturated, i.e. its magnetising impedance is very large ($Z_{eg} = \infty$). With reference to Figure 6.12b and the assumptions about the magnetising impedances, the voltage across the relay is $V = I_{sh}(R_{lh} + R_{sh})$ and the current through the relay is $I_R = V/R_R$. If the relay resistance is high (i.e. $R_R >> R_{lh} + R_{sh}$) then $I_{sh} \approx I_{sg}$.

The spill current can be reduced to a value below the operating threshold of the relay by a suitable choice of R_R. The resistor that must be connected in series with the relay to increase the self-resistance of the relay to R_R is known as the stabilising resistor. A suitable choice will ensure stability during all external faults, irrespective of the magnetic history of the CTs, and permit sensitive operation on all internal faults.

6.5.2 High-impedance differential

It is normal to provide phase and earth fault protection for the generator stator using the high-impedance differential system shown in Figure 6.13a [1]. Protection for earth faults only is achieved using the restricted earth fault differential system shown in Figure 6.13b. The latter is normally only used when the individual phase connections are not available at the neutral end of the generator.

The voltage setting for a high-impedance relay is determined by calculating the maximum possible voltage drop across the lead resistance and CT wiring resistance that could occur during a close-up solid external fault. The minimum primary operating current is calculated by adding the sum of the secondary exciting currents of all the parallel connected current transformers, when operating at the relay setting voltage, to the relay minimum operating current and multiplying the result by the CT turns ratio. Once the minimum primary operating current has been calculated, the region of the winding that is protected for earth faults can be evaluated [1, 9].

Example: A generator is rated at 4MVA, 11kV and its subtransient reactance is $j0.2$ per unit based on the machine rating. All the current transformers are 200/5 class X with a knee point voltage of 100V, a magnetising current of 100mA at 29V and a winding resistance of 0.6Ω. The loop resistance of the leads used to connect each CT to the relay is 0.5Ω. Determine the value of the stabilising resistor necessary to ensure the relay does not operate during a through fault if the relay operating voltage is 25V at a current of 25mA [1, 9].

For an external fault at the generator terminals, $I = kV / (\sqrt{3} \times Z)$, where $Z = Z_{pu} \times kV_{base}^2 / MVA_{base}$. Hence, $I = 6350 / (0.2 \times 11^2 / 4.0) = 1050A$ primary $= 5.25 \times I_{nominal} = 26.25A$ secondary. Voltage across relay

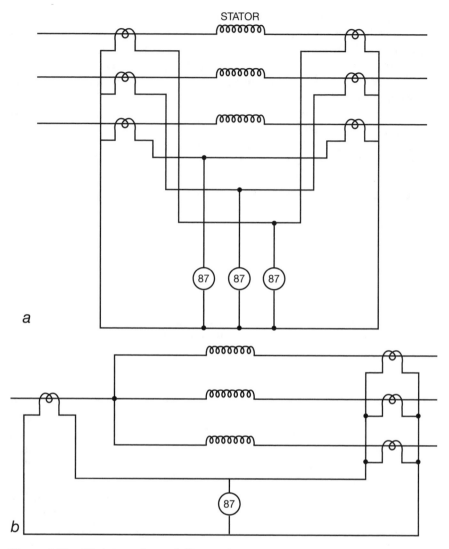

Figure 6.13   High-impedance differential protection:
a Phase and earth fault; b Restricted earth fault

and stabilising resistor = $26.25 \times (0.6 + 0.5) = 28.9\text{V}$. The operating voltage of the relay is 25V, and therefore 3.9V is dropped across the stabilising resistor. Since, the operating current is 25mA, the resistance of the stabilising resistor must be greater than 156 if stability is to be guaranteed on a close-up external fault.

If the neutral of the generator is earthed using a single-phase 6.3kV/240V distribution transformer with a secondary burden resistance of

0.1Ω, determine the fault current if an earth fault occurs at the line terminals of the stator winding and within the differential protection operating zone.

The impedance of the secondary burden resistance (0.1Ω) referred to the primary is 68.9Ω. Consequently, the current flowing into an earth fault at the terminals is $I = (11kV/\sqrt{3})/68.9\Omega = 92A$ primary.

The minimum primary operating current for a high-impedance differential scheme is POC = (excitation current of CTs in relay circuit plus operating current of the relay) × CT ratio.

Each relay in a phase and earth fault differential protection scheme (see Figure 6.13*a*), is connected to two CTs, and consequently the POC = $(2 \times 100mA + 25mA) \times (200/5) = 9A$. The POC for a restricted earth fault scheme (see Figure 6.13*b*), is $(4 \times 100mA + 25mA) \times (200/5) = 17A$.

To detect a fault on the stator winding the fault current must be greater than the minimum primary operating current. For an earth fault the fault current equals the per unit distance along the winding multiplied by the fault current caused by a terminal fault (example = 92A). Percentage of winding protected by phase and earth fault differential = $(1 - 9/92) \times 100\% = 90\%$. Percentage of winding protected by restricted earth fault differential = $(1 - 17/92) \times 100\% = 81\%$.

6.5.3 Low-impedance biased differential protection

Biased differential relays (87) are also used for phase fault protection of the stator winding [1, 11]. Phase-to-phase or three-phase faults normally result in high fault currents, and fast fault clearance is required. Some phase-to-earth faults may be detected but this depends on how the generator neutral is connected to earth. Traditional biased differential relays, when applied to a generator, would normally use a variable slope bias characteristic with the slope varying continuously from perhaps 5% at low values of through current to perhaps 100% at high currents. More recent designs tend to use a dual slope characteristic as shown in Figure 6.14. A typical value for the minimum current setting (I_{S1}) is $0.07I_n$, the low current bias (K1) is 10%, the threshold for increasing the bias (I_{S2}) is $2I_n$ and the high current bias (K2) is 50%, where I_n is the nominal current. The characteristic is designed to ensure that the lower slope provides high sensitivity for internal faults and the upper slope high stability for high current through faults that may result in saturation of a current transformer.

Bias is often used in conjunction with a stabilising resistor. This results in a differential protection scheme that is stable for all external faults, irrespective of the magnitude of the through current. Bias allows the use of a low ohmic value stabilising resistor. Consequently the minimum relay operating voltage is small and the relay achieves high sensitivity on internal faults.

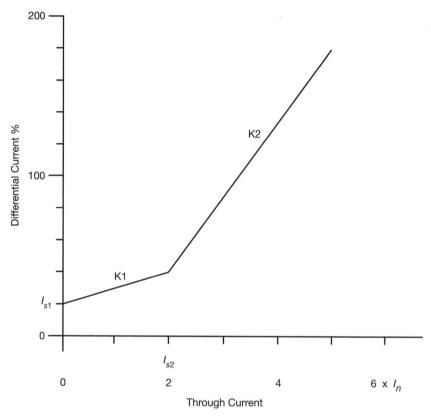

Figure 6.14   Operating characteristic for a biased differential relay used to protect the stator winding

6.6 Phase and interturn faults on the stator windings

A phase-phase fault on the stator winding that remains clear of earth is rare. They can occur on the end portions of the stator coils or in the slots if the winding involves two coil sides in the same slot. Particularly with the latter the fault will rapidly evolve to the core, i.e. to the earth. The current in a phase-phase fault is not limited by the earthing impedance, and consequently the fault current is high and the resulting damage, if the fault is allowed to persist, will be considerable.

An interturn fault, i.e. a fault that involves one/more turns of a winding and remains clear of the core, is also rare. Conventional protection cannot detect an interturn fault and the use of special systems for interturn fault detection cannot normally be justified due to their cost and complexity. Conventional protection will only detect this type of fault if it develops to the core; the resulting earth fault can then be cleared by overcurrent or overvoltage protection in the neutral.

6.7 Under/overvoltage protection

Many industrial networks operate with part of their electrical load supplied by a local generator with the remainder supplied by an intertie to the utility network [2, 3]. An undervoltage relay on this intertie will detect the voltage depression caused by a close-up utility fault and will then activate the tripping of the intertie circuit breaker. The industrial network will continue to be supplied by the local generator, although this may require the shedding of non-essential loads.

With a synchronous generator the automatic voltage regulator will normally maintain the terminal voltages of a generator within a narrow range of values, and consequently a sustained undervoltage indicates a severe overload condition or the loss of a generator. If a standby generator is available, an undervoltage relay with a long operating time can be used to initiate the start-up of the machine. Alternatively the undervoltage can be used to initiate the shedding of non-critical loads.

Overvoltage protection is normally provided on hydro-generators where excessive terminal voltages may occur immediately after load rejection. Sustained overvoltages may result in damage to the machine. With a hydro-generator, volts per hertz protection may not detect the problem, since the overvoltage will result in an overspeed and consequently the V/Hz operating settings may not be exceeded. In general, overvoltages are not a problem with steam or gas turbines since the rapid response of the speed governor and the voltage regulator prevents the occurrence. In addition, most excitation systems on synchronous generators have integral V/Hz limiters that prevent the overvoltage occurring.

An earth fault on a feeder connected to an unearthed transformer should be cleared by the feeder circuit breaker. However, if there is also a source of supply on the low-voltage side of the transformer, the feeder and the fault may remain energised. A voltage relay that measures the zero-sequence voltage on the feeder, i.e. the neutral displacement, can detect the fault. After a time delay designed to ensure adequate coordination with other relays on the system, the neutral displacement element can trip the circuit breaker on the low-voltage side of the transformer.

6.8 Under/overfrequency protection

Underfrequency relays are used to detect overloading of an embedded generator due to the partial or total loss of the grid supply to the local network. Underfrequency relays will normally trip the circuit breaker on the intertie between the generator and the utility network if the power frequency drops below a user-defined setting (typically 47–49Hz) for a time greater than a user-defined delay (typically 0.5–1.0s). The frequency setting should be set as low as possible since most utility distribution networks include underfrequency load shedding schemes that iteratively disconnect blocks of load as the frequency drops below a sequence of preset values. In the UK, the lowest preset value is 47Hz, i.e. all the load that can be shed will have been shed if the frequency collapse reaches 47Hz. Above this value it is still possible that sufficient load can be shed and the frequency will recover to 50Hz. Underfrequency tripping of embedded generators at frequencies above 47Hz reduces the probability of the national network successfully recovering from the initial overload.

Overfrequency relays are used to prevent damage to a generator caused by overspeeding resulting from a loss of load. Normally a speed control governor prevents overspeed and the overfrequency relay is only used as a back-up. The generator circuit breaker will be tripped if the power frequency increases above a user-defined setting (typically 50.5–52Hz) for a time greater than a user-defined delay (typically 0.5–1.0s).

6.9 Reverse power relay

The reverse power relay (32) provides back-up protection for the prime mover [1, 5]. It detects the reverse flow of power that results if the prime mover loses its source of input energy and the main generator circuit breaker fails to trip. The generator would start to motor, drawing real power from the utility network. A steam turbine could overheat due to the loss of the cooling effect of steam, a hydro-generator could suffer cavitation of the blades on low water flow, a diesel engine could either catch fire or explode due to unburned fuel and a gas turbine may have gearing problems when being driven from the generator end. Motoring protection must be provided for all generating units except those designed for operation as synchronous condensers. From an electrical viewpoint, the primary indication of motoring is the flow of real power from the utility network into the generator. The generator is behaving as a synchronous motor. A reverse power relay provides back-up for the mechanical devices that normally detect a motoring condition.

The settings and operating time delay applied to a reverse power relay depend on the type of prime mover used to drive the generator. For example, the compressor in a gas turbine has a substantial power requirement, up to 50% of the name-plate rating of the generator. Consequently, a low-sensitivity setting is required for the reverse power relay. Similarly a low-sensitivity setting is required with a diesel engine since with no cylinders firing the load is typically 25% of the generator rating.

With a hydro-turbine and a low water flow level the power required for motoring is only about 1% of rated power. Consequently the reverse power relay must be set very sensitively to detect the motoring condition. Similarly, steam turbines operating with full vacuum and zero steam input require about 2% of the rated power to motor. Again a sensitive reverse power relay is required.

The recommended setting for a reverse power relay is 10–20% of the maximum allowable motoring power and the operating time delay is typically 10–30 s. The time delay is required to prevent mal-operation during power swings or when synchronising the generator onto the network.

6.10 Loss of excitation

Loss of excitation can result from a loss of field to the main exciter, accidental tripping of the field breaker, short circuits in the field winding, inadequate brush contact in the exciter and loss of AC supply to the excitation system [5, 3, 1].

The effect of a loss of excitation on the impedance measured at the line terminals is shown in Figure 6.15. When the excitation fails, reactive power is supplied from the utility network, which causes the measured impedance value to move from quadrant 1 to quadrant 4 or from quadrant 4 towards quadrant 3. If the transient resulting from the loss of excitation continues, the trajectory of the measured impedance moves into the operating characteristic of the field failure relay. If the trajectory remains in the operating characteristic for a time greater than a user-defined setting, typically a few seconds, and if the terminal voltages are below the operating threshold for the undervoltage relays, then the generator circuit breaker is tripped. If the reductions in the voltages are insufficient to operate the undervoltage relay, operation of the field failure relay would normally activate an alarm. An operator would be expected to restore the field or initiate a controlled shutdown. In an unmanned station automatic time delayed shut-down of the machine may be initiated. Generally, generators must be kept on line as long as possible. In manned stations the operator must be given every opportunity to restore excitation and hopefully avoid unwanted tripping.

Field failure relays are designed to alert the operator to any reduction in excitation that may result in instability. The operator or field failure relay must only trip the generator when instability is imminent. Normally immediate tripping of the machine is only required if the terminal voltages are significantly below their nominal value and this threatens the

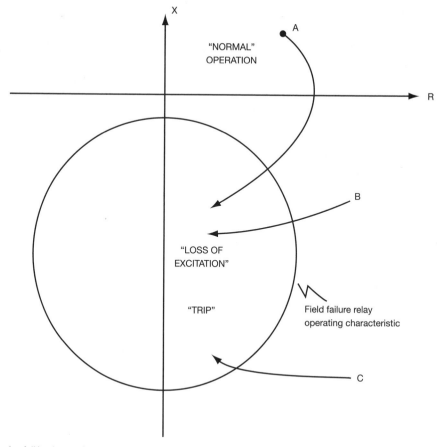

A = full load operating condition of generator
B = moderately loaded operating condition
C = lightly loaded operating condition

Figure 6.15   Effect on measured impedance of a loss of excitation

stability of the network. Generally, larger machines require several seconds to lose synchronism whilst smaller machines require only a few seconds.

6.11 Unbalanced loading

A three-phase balanced load results in a reaction field in the generator which is approximately constant and rotates synchronously with the rotor field system. Any imbalance in the load can be resolved into positive, negative and zero sequence components. A positive sequence effectively describes the components that are balanced in the same way as the normal balanced load. A zero sequence does not produce a reaction field. A negative sequence is similar to a positive sequence except the resulting reaction field rotates in the opposite direction to the field produced by the DC current in the rotor. The negative sequence flux cuts the rotor winding at twice the rotational speed and induces double-frequency currents in the field system and the rotor core. The resulting eddy currents in the rotor core tend to flow near the surface and in the non-magnetic rotor wedges and retaining rings. The resulting I^2R loss quickly raises the temperature, and if the imbalance persists the metal will melt, thus damaging the rotor structure [5, 3].

Unbalanced loads, unbalanced faults on the utility network, broken conductors or other asymmetrical operating conditions result in the unbalanced operation of a generator, the flow of negative sequence currents, overheating of the rotor and, if the condition is allowed to persist, serious damage to the machine. Most generators rated at less than 100MVA are able to operate continuously with negative sequence current imbalance of up to 10% of the full load current. If it is considered necessary to protect against imbalance, a negative sequence overcurrent relay can be applied.

6.12 Generator stator thermal protection

Thermal protection for the generator stator core and the windings is required to prevent damage due to generator overload, failure of the cooling system or localised hot spots caused by core lamination insulation failures or winding failures. Generally overheating is a long-term phenomenon that is not easily detected by conventional protection relays. In manned stations, thermal protection would normally activate an alarm. The operator then decides whether to trip the machine or take the steps necessary to reduce the temperature. In unmanned stations, thermal protection trips the machine.

Resistance temperature detectors embedded in the stator winding are normally used to sense the winding temperature. These detectors are monitored by a relay that operates if the resistance of any detector increases above a preset value. The value would normally correspond to a temperature of about 120°C.

If embedded temperature detectors are not available, a thermal replica relay can be used to model the effect on the stator temperature of an overload. The stator current is monitored and injected into a thermal replica that takes into account the thermal time constants of the machine and its previous thermal history. When the replica indicates that the temperature is in excess of the maximum allowable value for the insulation, the machine is tripped.

Overload protection can also be achieved using a time-delayed over-current relay, controlled by an instantaneous overcurrent element. For example, an extremely inverse overcurrent relay with a pick-up setting of 80% of full load current could be controlled by an instantaneous element with a pick-up setting of 120% of full load current. The relay is prevented from tripping for overloads below 120% of full load, but operates in a time defined by the extremely inverse characteristic for overloads above 120%. When determining the relay settings it is necessary to ensure that the overload protection does not mal-operate for a fault on the utility network.

Depending on the rating and type of generator the windings and stator core may be cooled by air, oil, hydrogen or water. In directly cooled generators, the coolant flows through the copper pipe that is the stator winding. In indirectly cooled generators, the coolant flows through pipes within the stator core or with small machines uses air flow over the surface of the core. Consequently, this relies on heat transfer through the winding insulation. A failure of the cooling system results in rapid deterioration of the winding insulation and the stator core lamination insulation. Most generator manufacturers provide protection for the cooling system. This involves sensors that monitor the winding temperature, the core temperature and the coolant pressure, temperature and flow. These sensors may be connected to an alarm or to a control system for automatically reducing the load or to a protection system that trips the main circuit breaker.

6.13 Overexcitation

Generators are designed to operate continuously at rated MVA, frequency and power factor and at a terminal voltage between 0.95 and 1.05 p.u. Overexcitation will occur if the ratio of voltage to frequency (volts/Hz) exceeds 1.05 p.u. This will result in saturation of the magnetic core of the generator and induced stray flux in non-laminated components which are not designed to carry flux. This causes severe overheating and eventually breakdown of the insulation [5, 3, 1].

The main cause of overexcitation is using the voltage regulator to control the operation of the generator at reduced frequencies. This can occur during start-up and shut-down. For example, if the generator is operating at a speed less than 95% of its rated value and the regulator is maintaining rated terminal voltage, then the volts/Hz will be greater than 1.05 p.u. This can result in thermal damage to the machine.

A volts/Hz limiter can be used to limit the output of the generator to a maximum volts/Hz value irrespective of the speed of the machine. The limiter operates in conjunction with the voltage regulator and limits the output voltage when the speed is low. Volts/Hz protection relays are designed to trip the generator if the volts/Hz exceeds a set value (typically 110% of its normal value) for a period greater than several seconds.

6.14 Loss of mains protection

Faults on a utility distribution network are normally cleared by relays located close to the fault. This can result in an embedded generator supplying part of the distribution network that has been islanded from the normal grid supplies. In most situations, the risk of the embedded generator continuing to operate without a grid connection is low, but it is not zero. The risk of a generator continuing to operate becomes greater when the generator or group of generators connected to the islanded network is/are capable of supplying all of its load. For safety reasons the protection applied to the intertie between the embedded generator and the utility network must be capable of detecting the loss of the grid supply, normally referred to as 'loss of mains' and tripping the circuit breaker on the intertie [10, 11, 12].

Islanding of an embedded generator may leave a section of the utility network without an earth and the fault level may be inadequate to adequately operate protection relays. In addition, circuit breakers on a distribution network do not normally include check synchronising facilities and consequently manual or automatic reclosure of the circuit breakers that isolate the islanded network from the grid supply could result in their destruction. The UK G59 recommendations state that immediately after the removal of the grid supply from a section of the utility network all the embedded generators connected to this islanded section must be automatically disconnected and remain disconnected until the normal grid supplies are restored. A loss of mains detection system is required in the UK for all MV connected generators and all LV connected generators operating with a rating greater than 150kVA.

Two techniques are normally used to detect loss of mains: rate of change of frequency (ROCOF) and vector shift.

6.14.1 Rate of change of frequency

ROCOF depends on the assumption that most 11kV utility distribution feeders are loaded up to about 5MW and following a loss of grid supply some of this load will have to be supplied by the embedded generator [8, 11, 12, 13]. The resulting generation deficit will cause a rate of change of frequency which, neglecting governor action, can be approximated by the following equation:

$$df/dt = -(P_{LO} \times f_r^2 - P_{TO} \times f_r^2) / (2 \times H \times P_{TNOM} \times f_r)$$

where P_{LO} = load in MW at rated frequency f_r
          P_{TO} = output of generating plant in MW
          P_{TNOM} = rated capacity of generating plant in MW
          H = inertia constant of generating plant in MW seconds
                     per MVA
          f_r = rated frequency.

For example, consider an 800kW 1MVA generator with an inertia constant H of 1.2s operating at 615kW when the loss of the grid supply occurs. If the load on the generator immediately increases to 677kW the initial rate of change of frequency is

$$df/dt = -(0.677 \times 50^2 - 0.615 \times 50^2) / (2 \times 1.2 \times 0.800 \times 50) = -1.6 \text{Hz/s}$$

The increase in load due to the loss of grid supply is 62kW which, assuming 5MW loading on the distribution feeder, is just over 1% of the distribution feeder loading.

The operating threshold for a ROCOF relay is normally adjustable between 0.1Hz/s and 10Hz/s and the operating time is defined by the number of power frequency measuring periods over which the rate of change of frequency is calculated. The minimum number of measuring periods is normally two (40ms at 50Hz) and the maximum typically 100 (2s at 50Hz). In the UK, typical df/dt settings are between 0.1 and 1.0Hz/s and the operating time is between 0.2 and 0.5s.

ROCOF has proved to be a sensitive and dependable method of detecting loss of mains when the rate of frequency change is relatively slow. This occurs when the size of the embedded generator is closely matched to the size of the islanded load.

The stability of a ROCOF relay is often considered inadequate since nuisance trips can result from frequency excursions caused by loss of bulk generation on the national network or phase shifts resulting from faults and switching on the local network. The latter can result in the incorrect calculation of df/dt particularly when the duration of the measuring window is small. In the UK, it is considered that major loss of bulk generation and uncontrolled tripping of transmission lines could result in rates of change of frequency as high as 1Hz/s. This would be an extremely rare event, but events resulting in frequency changes of perhaps 0.2 Hz/s occur reasonably often. During such incidents, tripping of large amounts of embedded generation via ROCOF relays could further risk the integrity of the national network. Large frequency changes on the national network are reasonably rare and the majority of nuisance trips are related to voltage dips and phase shifts caused by faults on the distribution network adjacent to the embedded generator.

ROCOF is generally considered an appropriate technique for detecting loss of mains on a distribution network. It is not ideal for detecting loss of mains at the subtransmission level (132kV in the UK) due to the increased risk of nuisance tripping caused by faults and sudden loss of bulk generation on the transmission network [8, 11, 12, 13].

6.14.2 *Vector shift*

During normal operation the terminal voltage of an embedded synchronous generator (V_t) will lag the synchronous electromotive force (E_f) by the rotor displacement angle φ (see Figure 6.16), which is defined by the voltage difference between E_f and V_t, i.e. $\Delta V = I_1 \times jX_d$. If the grid supply is suddenly disconnected from the section of the network partially supplied by the embedded generator, the load on the generator increases and this causes a shift in the rotor displacement angle (see Figure 6.17). The terminal voltage jumps to a new value and the phase position changes as shown in Figure 6.18.

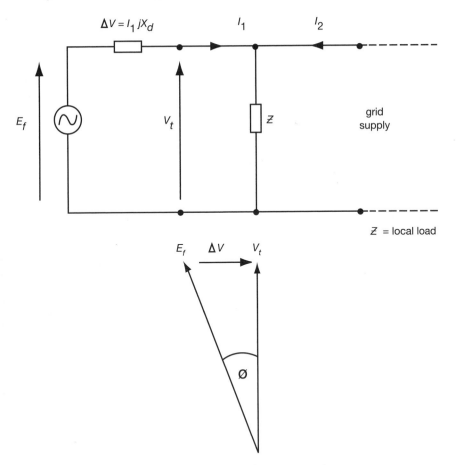

Figure 6.16   Normal operation – rotor displacement angle

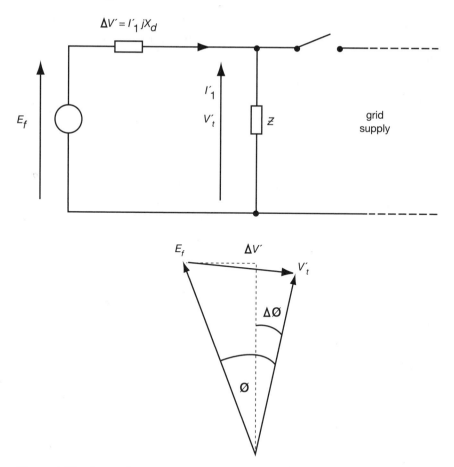

Figure 6.17   Loss of mains – rotor displacement angle

Vector shift relays continuously monitor the duration of each cycle and initiate instantaneous tripping if the duration of a cycle changes as compared to the previous cycle by an angle greater than the vector shift setting ($\Delta\Phi$). The recommended setting is 6°, but on weak networks it may be necessary to increase this to 12° to prevent mal-operation when switching on or off heavy consumer loads. In the UK typical vector shift settings vary between 8° and 12° [12, 13].

6.15 Rotor protection

Rotor protection is primarily concerned with the detection of earth faults in the field circuit of the generator. The field circuit consists of the rotor winding and the armature of the exciter and any associated circuit breaker. The field winding operates as an unearthed system and con-

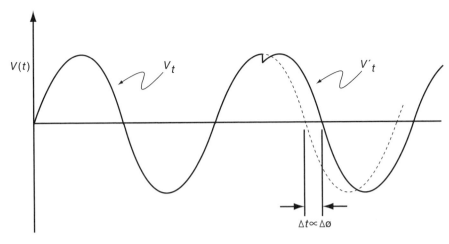

$\Delta t \propto \Delta \phi$

Figure 6.18   Voltage vector shift

sequently, for a fault between the field circuit and earth, no fault current flows and the generator continues to operate normally. However, if the fault was to develop to a second location, part of the field winding would be shorted, and unbalanced air-gap fluxes would be produced in the machine. These unbalanced fluxes could cause rotor vibration and damage to the machine. In addition, the current that flows in the shorted turns can result in localised heating and thermal damage to the winding. Generally, the probability of a second earth fault is greater than the first, since the initial fault establishes an earth loop for the voltages induced in the field by the current in the stator and this increases the voltage stress at other points on the field winding. Three methods are used to detect a rotor earth fault: the potentiometer method, the AC injection method and the DC injection method [1, 3, 5].

The potentiometer method consists of a centre tapped resistor connected across the field winding (see Figure 6.19). The centre point of the resistor is connected to earth through a voltage relay. Normally, the centre point is at 0V and the voltage across the relay is zero. During an earth fault, the centre point voltage changes, and the voltage relay detects the change and activates an alarm. The voltage change is a maximum for a fault at the ends of the winding and is zero for a fault at the electrical centre of the winding. To avoid the mid-winding blind spot, the tapping point must be regularly changed: modern designs achieve this automatically, but traditional designs rely on manual operation of a switch. To minimise the size of the blind spot, a sensitive setting is applied to the voltage relay, typically 5% of the excitation voltage.

The AC injection method (see Figure 6.20), involves capacitively coupling an AC voltage to the field circuit and measuring the AC current that flows to earth using a sensitive overcurrent relay (64). Under normal

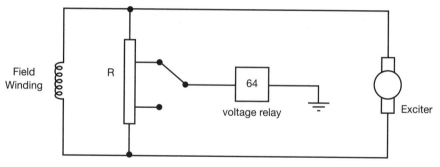

Figure 6.19   Earth fault protection of field circuit using potentiometer method

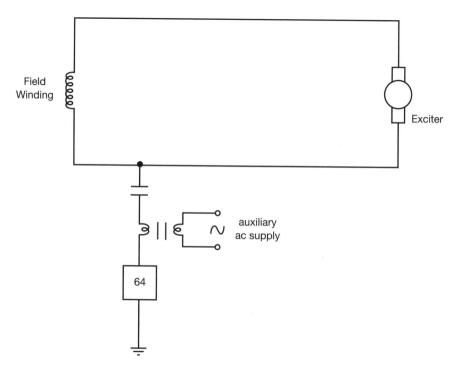

Figure 6.20   Earth fault protection of field circuit by AC injection

conditions, i.e. no earth fault, the only current that flows through the relay is the current that flows to earth through the capacitance of the winding. During an earth fault the current increases significantly, and the relay operates and activates an alarm. The main limitation of the AC injection method is that the capacitive current flows through the rotor bearings to earth and this may result in erosion of the bearing surface. This can be avoided by insulating the bearings and connecting an earthing brush to the rotor shaft.

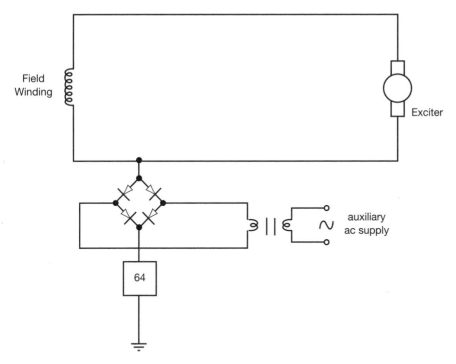

Figure 6.21   Earth fault protection of field circuit by DC injection

The DC injection method involves biasing the positive side of the field circuit to a negative voltage relative to earth (see Figure 6.21). The negative potentials of all points on the winding are more negative than the bias so an earth fault will cause current to flow through the sensitive overcurrent relay (64) connected in the earth connection. The impedance of the relay must be sufficient to limit the earth fault current to a value which cannot damage the winding. However, the settings must be designed to ensure that the relay is capable of detecting a high-resistance fault at all locations on the winding, whilst remaining stable during normal operating conditions.

6.16 References

1 'Protective relays application guide, 3rd edition, 1987'. Alstom T&D Protection & Control, St Leonards Works, Stafford ST17 4LX, UK
2 IEEE Recommended Practise for Protection and Coordination of Industrial and Commercial Power Systems, Buff Book, IEEE Std 242–1986, IEEE,ISBN 471–85392–5
3 ELMORE, W.A. (Ed.): 'Protective relaying theory and applications' (ABB Company, Marcel Dekker, 1994)
4 'IEEE guides and standards for protective relaying systems'. IEEE, 1991

5   'IEEE guide for AC generator protection'. IEEE C37.102, 1987
6   IEE Colloquium on *System Implications of Embedded Generation and its Protection and Control*, UK ISSN 0963–3308, IEE Digest No. 1998/277
7   ZIEGLER, G.: 'Protection of distributed generation – Current practice'. Symposium Neptun 1997, CIGRE 300–08, SC 34
8   FIELDING, G.: 'Protection of industrial power systems'. Lecture notes for MSc course on power system protection, UMIST, 1997
9   CROSSLEY, P.A.: 'Protection of generators'. Lecture notes for MSc course on power system protection, UMIST, 1998
10 'G59/1, Recommendations for the connection of embedded generating plant to the Regional Electricity Companies Distribution Systems'. Electricity Association, 1991
11 'Type LGPG111, Digital integrated generator protection relay'. Publication R4106B, Alstom T&D, Protection & Control
12 CROMPTON INSTRUMENTS: 'Vector shift and ROCOF relay', SW-ROCOF Edition 2. Crompton Instruments, Freebournes Road, Witham, Essex CM8 3AH, UK
13 'Mains protection by means of modern microprocessor technology'. SEG publicity pamphlet available from Schaltanlagen-Elektronik-Gerate, D-47884, Kempton 1, Postfach 10 07 67, Germany

Oxford eco house clad with BP Solar (high efficiency) photovoltaic panels. The 4 kWp array provides an average energy generation of 3.2 MWh per annum.

Photo: Peter Durrant

Chapter 7

Reliability concepts and assessment

7.1 Introduction

This chapter addresses the issues relating to the reliability assessment of distribution systems in which local generation is embedded. Superficially it could be expected that conventional approaches used for existing systems would suffice. However, this would be an erroneous conclusion for reasons discussed later in this chapter, although the required techniques can be based on developments of existing approaches. To appreciate all the issues involved it is therefore desirable to review existing approaches and their limitations regarding embedded generation before developing approaches and techniques to deal with these systems.

Reliability has always been an important system issue and it has been incumbent on power system managers, designers, planners and operators to ensure that customers receive adequate and secure supplies within reasonable economic constraints. Historically, this has been assessed using deterministic criteria, techniques and indices. As an example, the UK criteria centre on the application of Engineering Recommendation P2/5 [1]. This sets restoration requirements depending on the maximum demand of the load group being considered. In addition, the UK regional electricity companies (RECs) are expected to conform to sets of Guaranteed Standards [2] of service with penalty payments having to be made if these are violated. One Guaranteed Standard is service must now be restored within 18 h. Similar penalty payments are made in other countries such as Norway. The introduction of penalty payments indicates that economic methods for reducing the number of violations are very worthy of consideration.

As a concept, reliability is an inherent characteristic and a specific measure that describes the ability of a system to perform its intended function. In the case of a power system, the primary technical function is to supply electrical energy to its end customers (consumers). In the days of global, completely integrated and/or nationalised industries, the only significant measures required were those that were able to assess this overall function. However, in systems that are disaggregated, and in particular those that are owned or operated by private industry, there is also a fundamental need for all the individual parties (generators, network owners, network operators, energy suppliers) to know the quality of the system sector or subsector for which they (and their shareholders) are responsible. Therefore there is now a need for a range of reliability measures, the actual measure being dependent on whether it is for the use of generators, network owners, network operators, energy suppliers or the actual end customers.

Several issues affecting these different parties must be considered in the development of appropriate reliability evaluation models and techniques. These include the following. It is not necessary for the owner of the embedded generation to be the owner of the interfacing network. The transmission network and conventional generators will be affected as the combined installed capacity of embedded generation in the distribution network increases to significant levels. The load profiles seen at the bulk supply points will change and so will the power flows in both the transmission and distribution networks. Conventional generators will sell less energy and might be forced to operate outside optimum points because of the necessary load regulation action needed to compensate for random fluctuations in the output of the embedded generation. Therefore at least five different parties can be identified as being affected by the inclusion of embedded generation in distribution networks: the owner of the embedded generation, the interfacing distribution network, the customers, the transmission network and the conventional generators. Their perspectives are different and so are their interests and the benefits they derive from the locally generated energy. Hence the information required by each of the parties involved is different.

Assessing power system reliability is not a new activity, and a continuous stream of relevant papers have been published since the 1930s [3–8], with a selection of the most distinctive papers appearing in Reference 9. An extremely important aspect is that reliability levels are interdependent with economics [10] since increased investment is necessary to achieve increased reliability or even to maintain reliability at current and acceptable levels. This concept is illustrated in Figure 7.1, which shows the change in incremental cost of reliability ΔR with the investment cost needed to achieve it ΔC. It is therefore important to recognise that reliability and economics must be treated together in order to perform objective cost–benefit studies.

To review the functionality of a power system and the way this reacts with reliability, it is widely accepted internationally that the functional zones of a power system (generation, transmission and distribution) can be divided [11] into the three hierarchical levels shown in Figure 7.2. The

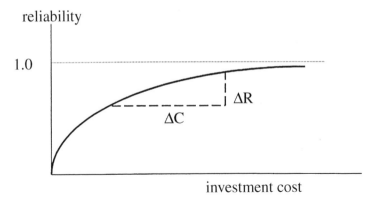

Figure 7.1 Incremental cost of reliability

first level (HLI) relates to generation facilities, the second level (HLII) refers to the integration of generation and transmission, and the third level (HLIII) refers to the complete system including distribution. This structure applied for several decades following the integration of the generation entity into large-scale and frequently remote sources. However, this relatively simple structure is now having to be reassessed due to two effects:

- the division of generation between a number of independent companies
- the inclusion of small-scale generation sources within the distribution system, forming embedded (or dispersed) generation.

7.2 HLI – generation capacity

During the planning phase of global generation, it was necessary to determine both how much capacity needed to be installed and when, so as to satisfy the expected demand at some point of time in the future and to provide sufficient reserve to perform corrective and preventive maintenance. This forms the HLI assessment. At HLI, the ability to move the energy to bulk supply points (BSPs) or end customers was not considered. Historically, the reserve capacity was set equal to a percentage of the expected load, or equal to one or more largest units, or a combination of both. These deterministic criteria could be, and in many cases were, replaced by probabilistic methods that responded to the stochastic factors influencing the reliability of the system.

A number of probabilistic criteria and indices are available for HLI studies [12–15]. These include loss of load probability (LOLP), loss of load expectation (LOLE), loss of energy expectation (LOEE), expected

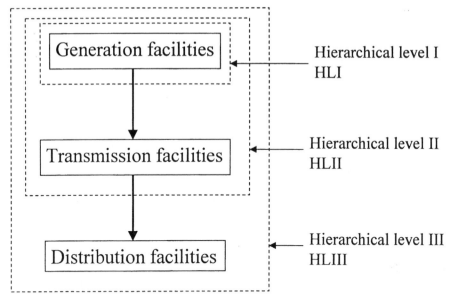

Figure 7.2   Hierarchical levels

energy not supplied (EENS), expected unserved energy (EUE), energy index of reliability (EIR), energy index of unreliability (EIU) and system minutes (SM). A brief description of each is given in Table 7.1. It is important to recognise that these indices are not measures actually seen by the end customers but are really relative measures by which alternative planning scenarios can be objectively compared.

Two scenarios can be studied at HLI depending on which energy sources are included in the assessments: first, by neglecting the generation sources embedded in the distribution system, and secondly by including them. Since HLI studies ignore the network completely and since embedded generation contributes to the energy supply, it seems appropriate to include such sources in HLI assessments. Studies of this type have been done (see Section 7.6). The problems of doing so are that, at the present time, they are generally not scheduled by system dispatchers, some types (e.g. wind, solar) are not easily predicted, and frequently they are not permitted to generate if they become disconnected from a BSP. Therefore their outputs are not easily incorporated into the assessments because of these significant dependency factors. Also their present contribution to the total system demand is relatively small. For all of these reasons, although it is wise and probably essential to include embedded generation in some way when making long-term strategic assessments, it may be preferable to neglect such sources when performing normal or routine HLI studies. However, as the

Table 7.1 HLI probabilistic criteria and indices

Index	Definition
LOLP	the probability that the load will exceed the available generation; it defines the likelihood of encountering trouble but not the severity
LOLE	the average number of days on which the daily peak load is expected to exceed the available generating capacity; alternatively it may be the average number of hours for which the load is expected to exceed the available capacity – again it defines likelihood, not severity
LOEE	the expected energy that will not be supplied due to those occasions when the load exceeds the available generation; it encompasses the severity of the deficiencies as well as their likelihood – essentially the same as EENS, EUE or similar terms
EIU	LOEE normalised by dividing by the total energy demanded
EIR	1-EIU
System minutes	LOEE normalised by dividing by the peak demand

latter are now decreasingly performed, the situation is probably of less significance.

7.3 HLII – composite generation and transmission systems

The second hierarchical level (HLII) is frequently referred to as a composite (generation and transmission) system or a bulk power (or transmission) system. The purpose of assessing the reliability of power systems at this level is to estimate the ability of the system to perform its function of moving the energy provided by the generation system to the BSPs. The distribution networks are not considered at this level.

Assessment of composite system reliability is very complex [12–16] since it must consider the integrated reliability effects of generation and transmission. These two entities cannot be analysed separately at this level. To do so could create misleading results and conclusions. This does not mean they have to be owned by the same company, but it is essential that one body has the role of co-ordinating the planning and operation. In some countries, one company is responsible for both planning and operation, whilst in others the responsibilities are divided between a network owner and an independent system operator (ISO). This depends on how the system ownership is structured.

The function of a composite system is to produce electrical energy at the generation sources and then to move this energy to the BSPs. The ability to generate sufficient energy to satisfy the demands at the BSPs, and to transport it without violating the system operational constraints, can be measured by one or more reliability indices. In HLI planning studies, the primary concern in the reliability assessment is one of steady-state adequacy, i.e. whether the generation system is adequate to meet the demands imposed on it. The dynamic behaviour of the system is generally only of concern during the operational phase.

This concept of steady-state adequacy is incomplete and therefore insufficient when the transmission system is incorporated into the analysis. In this case, the bulk transmission facilities must not only provide adequate transmission capacity to ensure the demand is satisfied and that voltage, frequency and thermal limits are maintained, but must also be capable of maintaining stability following fault, switching and other transient disturbances. The transmission facilities must, therefore, satisfy both static and dynamic conditions. Static evaluation is known as adequacy. The ability of the system to respond to disturbances arising in the system is known as security. The reliability of a composite system is therefore a matter of combining the two categories of adequacy and security [11, 17].

It should be noted that, although assessment of security is of great importance, virtually all of the techniques available at this time relate to adequacy assessment: probabilistic security assessment is very much a research topic at the present time. However, even with this restriction, planning decisions can be improved considerably with the additional objective information derived from adequacy assessments.

If the BSP of a transmission system supplies a distribution network which does not have any embedded generation, then the demand seen by that BSP is simply the aggregated demand within the distribution system being supplied. However, if the distribution system contains embedded generation then, not only will the energy demanded from the BSP (and therefore the transmission network feeding it) be reduced by the energy generated within the distribution system, but also the load profile seen by the BSP will change. This can have profound consequences. Since the embedded generation may not always be available, the maximum system demand on the BSP must still be considered to be the maximum demand that would exist even without the embedded generation. However, since the embedded generation will contribute for much of the time, the average demand seen by the BSP will decrease. Therefore the skewness and the variance of the demand seen at the BSP will increase, which will impose increased, and potentially considerable, variation on the demand at the BSP and consequently on the power flows through the transmission system feeding the BSP.

Although this is a direct consequence that cannot be eliminated, its impact can be assessed using conventional approaches for conducting HLII assessments. To do this, it is necessary to quantify the actual load profile seen at the BSP. This can be an output calculation from a reliability assessment conducted on the distribution system. In brief, if a time sequential estimate is made of the output of the embedded generators and these are convolved with the time sequential demands of the consumers in the distribution system, then a time sequential estimate of the remaining demand on the BSP can be made. This modified load profile can then be used in the HLII reliability assessment, which will provide a way of:

- comparing the transmission system behaviour and performance before and after the inclusion of embedded generation
- estimating the impact of revised power flows in the transmission network
- indicating the effect of new forms of generation on the existing large scale generators.

The above procedure of using the output of a distribution system assessment as input into HLII assessments is a fundamentally different approach since previously the HL procedures were a top-down concept in which the output of upper HL levels was used as the input to lower levels, not vice versa. However, it does confirm the versatility of the HL concept. For this approach to work, however, it is evident that the output of embedded generation as well as customer loads must be predicted or forecasted. Also, since the loads act as sinks for the energy produced by the embedded generation, it suggests that the dependency, if any, between embedded generation output and load demand needs to be established. An example of this could be a dependency between the output of a wind turbine and high load demand because both are affected by wind speed: the output of a wind turbine and the wind chill factor both increase with wind speed.

Although HLII reliability assessments have not been done routinely in the past, the competitive nature of the present electricity supply industry and the extended use of embedded generation, particularly renewables, suggest that such assessments could be of increasing importance. The main reason is that competition must be conducted objectively to ensure that all parties involved in energy trading and scheduling are treated fairly and impartially. Also operational decisions should be subjected to technical and economic auditing, and quantitative reliability assessments are able to provide information to assist in such audits.

It is worth noting at this point that a distribution system containing dispersed generation is equivalent to a composite generation and transmission system, and therefore the techniques and models developed for HLII studies could conceivably be applicable to such distribution systems. In concept this is true and several approaches for assessing the reliability of systems with embedded generation are based on these approaches. However, it must also be noted that the assessment objective at HLII is different from that required at the distribution level: essentially the objective at HLII is to assess the ability to supply energy to BSPs (the actual consumer is not directly involved in the assessment), whereas the objective of a distribution system is to supply end customers. In consequence the objectives for distribution systems containing embedded generation is multifold.

7.4 HLIII – distribution systems without embedded generation

7.4.1 Conceptual requirements

Complete consideration of HLIII would enable the effect of generation, transmission and distribution on individual customers to be evaluated and compared against relevant design and operational criteria. Although this total effect can be monitored realistically in terms of past performance measures, it is usually impractical for future system predictions because of the enormity of the problem. Instead the predictive assessment is usually done for the distribution functional zone only. Traditionally this is acceptable because:

(a) distribution networks generally interface with the transmission system, through which the effects of global sources of generation are seen, via one supply point. Therefore the load point indices evaluated in the HLII assessments could be used as input values for the reliability evaluation of a distribution system to give overall HLIII indices. This concept becomes less applicable with embedded generation since these generation sources are dispersed throughout the network rather than being all focused at one point (BSP)

(b) distribution systems are generally the major cause for the outages seen by individual customers and therefore dominate the overall reliability indices: generally on a worldwide consideration 80–95% of customer unavailability can be accounted for by the distribution network. Typical results are shown in Figure 7.3. These represent the UK 10 year average values [18] for availability and security (see Section 7.4.2 for definition of these terms).

The technical function of a distribution system is to take energy from BSPs and deliver it to individual end customers within certain quality constraints of voltage, frequency, harmonics, flicker, etc. It is also expected to achieve this with a reasonable level of reliability, i.e. to keep the number and duration of outages reasonably low. This can be quite difficult to achieve economically, particularly at the lower voltage levels and in rural areas, because the system generally consists of single radial overhead lines that are exposed to adverse environmental conditions. They are therefore prone to failure and frequently lengthy outage times.

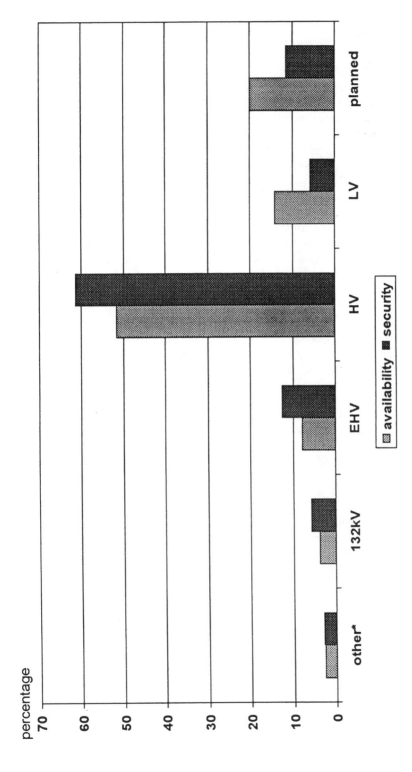

Figure 7.3 Typical contributions to customer reliability [2]
*includes generation and transmission

Superficially the inclusion of local sources of generation would seem to provide a remedy for reducing the number and duration of outages experienced by customers but, although this is always possible theoretically and could be so in some systems, in others it will not provide such immediate benefits because of other technical operational and safety constraints. These constraints are discussed in Chapter 1 and revisited briefly in Section 7.5.

The naturally occurring environmental conditions which enhance the failure process cannot be avoided, but the effects can be considered by assessing objectively the adequacy of the supply seen by customers. Such assessments are necessary for several reasons:

(a) Although a given reinforcement scheme may be relatively inexpensive, large sums of money are expended collectively on such systems.
(b) It is necessary to ensure a reasonable balance in the reliability of the various constituent parts of a power system, i.e. generation, transmission, distribution.
(c) A number of alternatives are available to distribution engineers to achieve acceptable customer reliability, including alternative reinforcement schemes, allocation of spares, improvements in maintenance policy, alternative operating policies, and maybe the option of embedded generation.

It is not possible to compare quantitatively the merits of such alternatives or to compare their effect per unit cost expended without utilising quantitative reliability evaluation. These assessments are relatively simple in distribution systems without embedded generation, particularly at the lower voltage levels because:

(a) the system is frequently either radial, or operated radially, and the evaluation techniques are very straightforward, often not requiring complex computer programs
(b) the complex problem of security which occurs at HLII does not exist at the distribution level and therefore steady-state adequacy assessment is sufficient.

These relatively simple approaches are not so amenable when embedded generation is dispersed throughout the distribution network. Two particular aspects are immediately apparent. First, the assumption of one single input source of generation is no longer valid since a number of dispersed sources may exist, and these may also have operational dependencies with each other and with the main BSP input. Secondly, with the inclusion of dispersed generation, distribution systems exhibit characteristics similar to those of transmission systems and the dynamic and security concepts discussed in Section 7.3 become applicable: the consideration of adequacy alone is therefore not acceptable.

7.4.2 Probabilistic criteria and indices

Most utilities collect measures of how distribution systems perform during the operational phase. Historically these are customer-related measures evaluated from system interruption data. The basic indices are failure rate λ, average outage duration r, and annual unavailability U at individual load points. A set of system indices can be deduced using customer and load data. The system indices listed below are generally accepted internationally. However, since two additional terms used by the UK regulator [18] are creeping into international usage, they are also included below.

- System Average Interruption Frequency Index (SAIFI)
    In the UK this is equivalent to the term, SECURITY, which is defined as the number of interruptions per 100 connected customers per year [18].
- System Average Interruption Duration Index (SAIDI)
    In the UK this is equivalent to the term, AVAILABILITY, which is defined as the number of customer minutes lost (CML) per connected customer per year [18].
- Customer Average Interruption Frequency Index (CAIFI)
    In the UK this is equivalent to the number of CML per interruption
- Customer Average Interruption Duration Index (CAIDI)
- Average Service Availability Index (ASAI)
- Average (Expected) Energy Not Supplied (AENS/EENS)

These indices are defined in Table 7.2 and are excellent measures for assessing how well a system has performed its basic function of satisfying the needs of its customers. The indices can be calculated for the overall system or for subsets of the system, e.g. individual feeders, service areas, etc., depending on the requirements for the performance measures.

Table 7.2   Definitions of system indices

Index	Description
SAIFI	average number of interruptions per customer served per year
CAIFI	average number of interruptions per customer affected per year
SAIDI	average interruption duration per customer served per year
CAIDI	average interruption duration per customer interruption
ASAI	ratio of the total number of customer hours that service was available during a year to the total customer hours demanded
ASUI	ratio of the total number of customer hours that service was unavailable during a year to the total customer hours demanded
AENS	average energy not supplied per customer served per year

The ability to calculate the same range of indices for future performance is an important consideration. However, it does require realistic component data that include relevant failure rates and restoration times. This is not easily obtained from some fault reporting schemes which record information only when customers are interrupted and not for equipment or component failures when customer outages do not occur. Consequently such data may be limited in predicting future system behaviour.

7.4.3 Historical evaluation techniques

Reliability assessment of distribution systems has received considerable attention and there are a large number of publications dealing with the theoretical developments and applications [3–9]. The usual method for evaluating the reliability indices is an analytical approach [12] based on a failure modes assessment and the use of equations for series and parallel networks [19]. A simulation approach is sometimes used for special purposes to determine, for instance, the probability distributions of the reliability indices [20]. The assessment procedure is to evaluate the reliability indices at each individual load point by identifying the events leading to failure of the load point and using appropriate equations to evaluate the indices.

Radial networks, or meshed ones operated radially, are the simplest to assess. In these cases the components are all in series and the equations needed to evaluate the basic indices are very simplistic, as shown in Figure 7.4a.

The process is more complex for parallel or meshed systems. In this case the failure modes of the load point involve overlapping outages, i.e. two or more components must be on outage at the same time (an overlapping outage) to interrupt the load point. Assuming that failures are independent and that restoration involves repair or replacement, the equations used to evaluate the indices of the overlapping outage are shown in Figure 7.4b. The indices λ_p and r_p then replace λ_i and r_i in the equations shown in Figure 7.4a to give the overall load point indices.

This approach is straightforward and only requires an understanding of the way in which each load point can fail and the relevant reliability data of the components leading to that load point failure. After evaluating the basic indices for each load point, the system indices, SAIFI, SAIDI (or SECURITY and AVAILABILITY in the UK), etc., can be evaluated using the previously described principles.

This basic approach relates to a process involving a single failure mode of a component, a single repair or replacement procedure and independent failures between components. Although this is generally sufficient for single-line radial networks, it frequently needs extending to cope with

$$\lambda_s = \Sigma \; \lambda_i$$

$$U_s = \Sigma \; \lambda_i r_i$$

$$r_s = U_s / \lambda_s$$

$$E_s = L.U_s$$

where λ_i is component failure rate

r_i is component restoration time

L is average load at the load point

(a) series components

$$\lambda_p = \lambda_i \lambda_j (r_i + r_j)$$

$$r_p = r_i r_j / (r_i + r_j)$$

(b) overlapping outages

Figure 7.4   Basic equations for reliability indices

more complex structures or when it is necessary to consider additional component failure/restoration processes. These additions (more details regarding techniques are available elsewhere [12]) include the following:

(a) weather effects. Weather and other adverse environmental conditions can significantly enhance the failure rate of equipment, particularly overhead lines. This enhancement greatly increases the likelihood of an overlapping outage of two or more components and causes a phenomenon known as 'bunching' of failures. The difficulty in predictive assessments is in ascertaining the relationship of failure rate with weather

(b) common mode failures. A common mode failure can occur when a single external event causes the simultaneous outage of two or more components. For it to be a common mode event, these simultaneous outages must be directly caused by the single external event and not be consequences of each other. These events are often difficult to identify but can have a significant impact on system behaviour

(c) types of outages. Each component of a system can fail or be outaged in a number of ways. It is often unrealistic to group or pool these together and associate with them a single average value of failure rate and restoration time. For instance, separation of events is important if they have different effects on the system or are associated with very different restoration processes. Four particular outage events are permanent, temporary, transient and scheduled, as defined in Table 7.3

(d) partial loss of continuity. The overlapping events so far discussed are considered to cause total interruption of the supply to a load point; this is known as total loss of continuity. However, a component outage in a meshed or looped system may cause network constraints to become violated, and some load, not all, is then disconnected at one or more load points. These additional failure events are said to cause partial loss of continuity (PLOC) [21]

(e) effect of transferrable loads. Many systems, although operated radially, are structured as meshed systems with open points. When an outage occurs it may be possible to relocate the normally open points and transfer some or all disconnected load to another supply point. Transfer restrictions may exist due to insufficient supply point capacity and/or feeder capacity

(f) embedded generation. As discussed in Section 7.1, the inclusion of embedded generation in the distribution network may superficially be expected to improve customer reliability. Whether it does so or not depends on system operational policies. As the main concern of this book, this aspect is discussed in more detail in Section 7.5 onwards.

7.4.4 Basic reliability assessments

It is not possible within the short space of this chapter to describe detailed sets of analyses. For this information the reader is referred to

Table 7.3   Definition of outage types

Permanent outages	associated with damaged faults requiring the component to be repaired or replaced
Temporary outages	associated with undamaged faults that are restored by manual switching or fuse replacement
Transient outages	associated with undamaged faults that are restored automatically
Scheduled maintenance outages	those outages which are planned in advance to perform preventive maintenance

Reference 12 or one of the many application papers [3–9]. Instead a few results based on those given in Reference 12 are used to illustrate some of the concepts described above.

Consider the radial system shown in Figure 7.5. The basic function can be achieved using solid feed-points and no isolators/disconnects. Such an arrangement would be perfectly adequate if no failures were to occur. This, however, is not realistic and protection devices and disconnects are usually installed in order to improve system reliability. Consider the situation of installing fuses F1–F4 and disconnects D1–D3 at the points shown in Figure 7.5. Also assume the reliability and loading data shown in Table 7.4. These data assume the demand at each load point remains constant, i.e. it represents the average load. This need not be a restriction since the analysis can be divided into appropriate time periods, the loading level in each time period determined from the load duration curve, reliability results obtained for each time period, and these individual results weighted together to provide overall values.

Using the equations shown in Figure 7.4 gives the results similar to those given in Reference 12 and shown in Table 7.5. These results include the individual failure events, the reliability indices for each load point, and some of the overall system indices. It is clearly evident from these results that the well known phenomenon of non-uniform customer reliability is encountered.

A similar set of results to those shown in Table 7.5 could be obtained from sensitivity studies that consider alternative arrangements of fuses and disconnects, as well as other forms of improvements such as back-feeding and possible use of embedded generation. Each alternative produces a change in the reliability indices with different customers benefiting by different amounts (some may not benefit at all in certain cases). These results can therefore be used as base case values so that the benefit (if any) of having embedded generation can be assessed. The comparisons are made in Section 7.7.1.

A subsequent question [10] then arises: 'Is the improvement worth it?'. This leads to a discussion of economics and customer evaluation of worth of supply, since the actual conclusion as to which is 'best' or

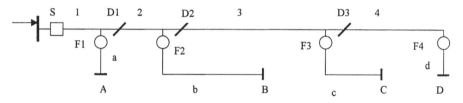

Figure 7.5   Radial distribution system

Table 7.4   System data (based on Reference 12)

1. Reliability data

Component Section	Failure rate, f/year	Repair time, h	Switching time, h
1	0.2	4	0.5
2	0.1	4	0.5
3	0.3	4	0.5
4	0.2	4	0.5
Lateral			
a	0.2	2	0.5
b	0.6	2	0.5
c	0.4	2	0.5
d	0.2	2	0.5

2. Loading data

Loading point	Number of customers	Total load connected, kW
A	2000	2500
B	1600	2000
C	1200	1500
D.	1200	1500

'worst' depends on the requirements of the customers, their expectations and outage costs. The use of cost–benefit analyses based on customers' evaluation of worth of supply is given in References 22–27.

7.5   Distribution systems with embedded generation

7.5.1   Concepts of embedded generation

It is not the purpose of this chapter to discuss purposes, objectives and types of embedded generation (this is discussed in Chapter 1). However, it is pertinent to reflect on these issues in respect of their impact on reliability, its evaluation and the models used. The installation of relatively small-scale generation, including non-utility sources, within a distribution network imposes additional modelling requirements. In fact the distribution network, historically mainly a supplier of energy received from bulk supply points, is becoming a mini-composite system having characteristics previously associated only with HLII.

The sources of generation considered in this book are generally small scale and therefore it is technically inappropriate for them to be con-

Table 7.5   Reliability indices for no embedded generation (based on reference 12)

Failure	Load point A			Load point B			Load point C			Load point D		
	λ	r	U	λ	r	U	λ	r	U	λ	r	U
Section												
1	0.2	4	0.8	0.2	4	0.8	0.2	4	0.8	0.2	4	0.8
2	0.1	0.5	0.05	0.1	4	0.4	0.1	4	0.4	0.1	4	0.4
3	0.3	0.5	0.15	0.3	0.5	0.15	0.3	4	1.2	0.3	4	1.2
4	0.2	0.5	0.1	0.2	0.5	0.1	0.2	0.5	0.1	0.2	4	0.8
Lateral												
a	0.2	2	0.4	–	–	–	–	–	–	–	–	–
b	–	–	–	0.6	2	1.2	–	–	–	–	–	–
c	–	–	–	–	–	–	0.4	2	0.8	–	–	–
d	–	–	–	–	–	–	–	–	–	0.2	2	0.4
Total	1.0	1.5	1.5	1.4	1.89	2.65	1.2	2.75	3.3	1.0	3.6	3.6

Security = 115 interruptions/100 customers
Availability = 155 min/year
Average duration = 2.25 h/interruption
Expected energy not supplied = 19 400 kWh or 3.23 kWh/customer
Energy demanded from BPS = 65.7 GWh

nected directly to the grid or transmission system. Instead they are generally connected to the distribution network with all the consequential technical problems that ensue. These are particularly the weak nature of the network in which they are embedded, and the fact that they exist much closer to actual consumers than large-scale global generation. All the reliability models and evaluation techniques must reflect this connection situation.

Furthermore, because these sources are dispersed around a weak network, they can frequently become disconnected from BSPs but still be connected to consumers. This leads to the question of whether they can be allowed to continue supplying load (much more preferable from a reliability point of view) or whether they must be tripped (preferable from a safety point of view). Both operational forms can be included in the reliability assessments.

7.5.2 Types and impact of energy sources

Embedded generation in theory could be sourced from many forms of primary energy. The current types are described in Chapter 1 and include wind, solar, CHP, small-scale hydro, biomass, land-fill gas, etc. However, from a reliability modelling and evaluation viewpoint, they can generally be grouped into two main types: those which have an output dependent on a variable energy source (e.g. wind, solar) and cannot be prescheduled even if it was wished to do so, and those that are not so dependent (e.g. hydro, gas, diesel) and could be prescheduled. The latter type can be modelled using conventional generation approaches [12–15] and their contribution to the system supply is only dependent on need and the availability of the units themselves. The former, however, are much more difficult to deal with because their contribution to the system supply is also dependent on the primary source of energy being available (if this energy source is quiescent then all units in the same geographical area are likely to be equally affected giving an outcome that could be considered equivalent to a common mode failure) as well as need and unit availability. Models and techniques for such intermittent sources are available and are discussed in Section 7.8.

At this point, it is useful to comment on CHP specifically. Since these sources can be controlled and the sources of primary energy are not dependent on variable environmental conditions, they could, and maybe should, be considered to contribute to the second type discussed above, namely the same as gas and diesel, for instance. However, this neglects the fact that they are also heat sources and the available electrical energy output is therefore dependent on other factors, e.g. the heat output and whether they are economical to run if heating is, or is not, required, etc. Since CHP sources can be controlled and therefore scheduled, it is reasonable to include such sources in the second type from a reliability point of view though not necessarily from an economic one. Alternatively these sources could be included in the first type and the output related to other environmental factors, e.g. temperature, in a similar way that the energy output of wind turbines is related to wind speed.

7.6 Historical reliability assessment approaches

Initial studies of the reliability impact of unconventional energy sources (generally renewables) have dealt with the problem only at the generation level (HLI) and have mainly centred on large-scale hydro, small-scale run-of-river hydro, and wind energy conversion systems. Very little has been done regarding generation embedded in distribution systems. The first attempts to include unconventional energy sources in reliability analysis of the generating system used the loss-of-load approach. A method which included frequency and duration concepts is presented in [28] for wind energy conversion systems. The method presented in [28] is enhanced to consider different types of units within the same wind farm [29, 30] and average wind velocity spatial correlation [30]. Another approach using the cumulant method including the failure and repair characteristic of the wind turbines is illustrated in [31]. Correlation between load and the power output of unconventional units is taken into account in [32]. Later publications use a loss-of-energy approach (LOEE). Methods based on the load modification technique were developed in [33, 34], with economic assessment included in [33]. None of these previous approaches consider explicit modelling of all the factors affecting the source of generation, e.g. wind farm, and instead treat the overall plant or farm as a single entity. The effects of the individual components and factors are therefore masked and the impact of being embedded in a weak network is generally neglected. However, some recent publications have appeared that have included these additional factors [35–39].

7.7 Simplified case studies

7.7.1 Basic radial systems

To illustrate the principles associated with the reliability assessment of distribution systems containing embedded generation, the system shown in Figure 7.5 is modified to include an embedded generation plant cited at the end of Section 7.4 as shown in Figure 7.6. Two main cases are considered: first, when the embedded generation can be operated independently of the main supply from the BSP and secondly, when it must be tripped following the failure of the supply from the BSP:

(a) Embedded generation cannot be run independently of the BSP: Consider first the situation when the embedded generation cannot be allowed to continue to operate following a disruption of supply from the BSP: this may occur due to a failure of the BSP, of the transmission system leading to the BSP, or of the distribution feeder between the BSP and the load centre being assessed. Furthermore, consider the case when the capability of the embedded generation exceeds that of the maximum demand of the system, i.e. greater than 7500kW, and is available whenever needed. The reliability results for this situation are shown in Table 7.6.

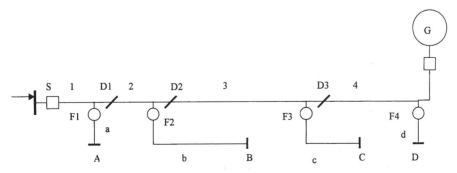

Figure 7.6   Radial distribution system with embedded generator

Table 7.6   Reliability indices with embedded generation (capacity >7500 kW, tripped when supply is interrupted)

Failure	Load point A			Load point B			Load point C			Load point D		
	λ	r	U	λ	r	U	λ	r	U	λ	r	U
Section												
1	0.2	4	0.8	0.2	4	0.8	0.2	4	0.8	0.2	4	0.8
2	0.1	0.5	0.05	0.1	4	0.4	0.1	4	0.4	0.1	4	0.4
3	0.3	0.5	0.15	0.3	0.5	0.15	0.3	4	1.2	0.3	4	1.2
4	0.2	0.5	0.1	0.2	0.5	0.1	0.2	0.5	0.1	0.2	4	0.8
Lateral												
a	0.2	2	0.4	–	–	–	–	–	–	–	–	–
b	–	–	–	0.6	2	1.2	–	–	–	–	–	–
c	–	–	–	–	–	–	0.4	2	0.8	–	–	–
d	–	–	–	–	–	–	–	–	–	0.2	2	0.4
Total	1.0	1.5	1.5	1.4	1.89	2.65	1.2	2.75	3.3	1.0	3.6	3.6

Security = 115 interruptions/100 customers
Availability = 155 min/year
Average duration = 2.25 h/interruption
Expected energy not supplied = 19 400 kWh or 3.23 kWh/customer
Energy demanded from BPS = 39.4 GWh
Energy delivered by embedded generation = 26.3 GWh, i.e. import reduction

It is seen that the load point and system indices shown in Table 7.6 are exactly the same as those in Table 7.5 for the case when there was no embedded generation. This can be expected for the specified operating conditions since, whenever a failure occurs, the embedded generating plant is also disconnected. The one distinctive difference is that the energy delivered by the BSP is reduced from 65.7GWh to 39.4GWh, a reduction of 26.3GWh, the energy delivered by the embedded generation plant. This confirms that one of the main contributions made by embedded generation is energy replacement. It could also contribute to a reduction of system losses, but this is not indicated in this example because it considers continuity only and therefore neglects power flows and the resulting effects.

This is an oversimplified example for several reasons. First, it assumes that the embedded generation is always available: this aspect is dealt with further in Section 7.8. Secondly, it neglects two situations which can occur in real systems when the embedded generation can contribute to increased reliability:

• If an outage occurs in a meshed, looped or parallel system, a partial loss of continuity (PLOC – see Section 7.4.3) event may occur without embedded generation but not when it exists. Such events are considered in the more comprehensive approach described in Section 7.9.

- Following year-on-year load growth, network capacity may be reached requiring reinforcement of the network. An alternative is to provide local generation which enables continuity of supply to be maintained even at peak demand. This aspect is considered briefly in Section 7.7.2.

(b) Embedded generation can be run independently of the BSP: If the embedded generation can be used independently of the main supply from the BSP, then this source can continue to supply all or some of the load when a failure occurs between the BSP and the load centre being assessed. This will mean an increase in the system reliability. Three case studies are included to illustrate this aspect:

- Embedded generation is always available and has a capability greater than the maximum demand.
- Embedded generation is always available but has a capability of 3000kW, which is less than the maximum demand.
- Embedded generation is available only for 25% of the time and has a capability of 3000kWh.

The results for the three case studies are shown in Tables 7.7–7.9.

Table 7.7 Reliability indices with embedded generation (capacity > 7500 kW, can run independently)

Failure	Load point A			Load point B			Load point C			Load point D		
	λ	r	U	λ	r	U	λ	r	U	λ	r	U
Section												
1	0.2	4	0.8	0.2	0.5	0.1	0.2	0.5	0.1	0.2	0.5	0.1
2	0.1	0.5	0.05	0.1	4	0.4	0.1	0.5	0.05	0.1	0.5	0.05
3	0.3	0.5	0.15	0.3	0.5	0.15	0.3	4	1.2	0.3	0.5	0.15
4	0.2	0.5	0.1	0.2	0.5	0.1	0.2	0.5	0.1	0.2	4	0.8
Lateral												
a	0.2	2	0.4	–	–	–	–	–	–	–	–	–
b	–	–	–	0.6	2	1.2	–	–	–	–	–	–
c	–	–	–	–	–	–	0.4	2	0.8	–	–	–
d	–	–	–	–	–	–	–	–	–	0.2	2	0.4
Total	1.0	1.5	1.5	1.4	1.39	1.95	1.2	1.88	2.25	1.0	1.5	1.5

Security = 115 interruptions/100 customers
Availability = 106.2 min/year
Average duration = 1.54 h/interruption
Expected energy not supplied = 13 275 kWh or 2.21 kWh/customer

Table 7.8   Reliability indices with embedded generation (capacity = 3000 kW available continuously, can run independently)

Failure	Load point A			Load point B			Load point C			Load point D		
	λ	r	U	λ	r	U	λ	r	U	λ	r	U
Section												
1	0.2	4	0.8	0.2	4	0.8	0.2	0.5	0.1	0.2	0.5	0.1
2	0.1	0.5	0.05	0.1	4	0.4	0.1	0.5	0.05	0.1	0.5	0.05
3	0.3	0.5	0.15	0.3	0.5	0.15	0.3	4	1.2	0.3	0.5	0.15
4	0.2	0.5	0.1	0.2	0.5	0.1	0.2	0.5	0.1	0.2	4	0.8
Lateral												
a	0.2	2	0.4	–	–	–	–	–	–	–	–	–
b	–	–	–	0.6	2	1.2	–	–	–	–	–	–
c	–	–	–	–	–	–	0.4	2	0.8	–	–	–
d	–	–	–	–	–	–	–	–	–	0.2	2	0.4
Total	1.0	1.5	1.5	1.4	1.89	2.65	1.2	1.88	2.25	1.0	1.5	1.5

Security = 115 interruptions/100 customers
Availability = 117.4 min/year
Average duration = 1.70 h/interruption
Expected energy not supplied = 14 675 kWh or 2.45 kWh/customer

Table 7.9   Reliability indices with embedded generation (maximum capacity of 3000 kW available only for 25% of time, can run independently)

Failure	Load point A			Load point B			Load point C			Load point D		
	λ	r	U	λ	r	U	λ	r	U	λ	r	U
Avail.	1.0	1.5	1.5	1.4	1.89	2.65	1.2	1.88	2.25	1.0	1.5	1.5
Unav.	1.0	1.5	1.5	1.4	1.89	2.65	1.2	2.75	3.3	1.0	3.6	3.6
Total	1.0	1.5	1.5	1.4	1.89	2.65	1.2	2.53	3.04	1.0	3.08	3.08

Security = 115 interruptions/100 customers
Availability = 145.8 min/year
Average duration = 2.11 h/interruption
Expected energy not supplied = 18 230 kWh or 3.04 kWh/customer

It can be observed in all cases that the failure rate, and therefore security, remain unchanged. This is logical because, when a fault occurs, the protection breakers at both the BSP and the embedded generating plant must be tripped. This is followed by opening appropriate disconnects and restarting the generator. An improvement in the failure rates and security could be achieved by using protection breakers in place of some or all the disconnects. (The reader should refer to Chapter 6 for more detail regarding protection schemes and procedures.)

However, it can also be observed that there is a significant reduction in the outage times, the annual unavailabilities and the overall availability. This is due entirely to the benefit of being able to pick up some loads by the embedded generator following switching while the main supply is being restored. This benefit is greatest in the case when the generator is always available and has a capability greater than the maximum demand (Table 7.7): the benefit decreases when the generator capacity is decreased to 3000kW (Table 7.8), and further decreases when the generator is available for only 25% of the time (Table 7.9). The first two situations would reflect usage of, say, diesel generators and the last situation would reflect usage of wind or solar, if it was permitted to use these independently of the main supply.

7.7.2 Embedded generation vs. network expansion

As discussed in Section 7.7.1, one possible benefit of embedded generation is to use it as a reinforcement strategy instead of enhancing the network following load growth. This is permissible under P2/5 [1]. To illustrate this aspect, consider the simple radial system feeding one load point shown in Figure 7.7. Assume the load is expected to grow from 10MW, the present rating of the network, to 12MW. Two possible reinforcement strategies are: to provide a new line in parallel with the existing one, or to install one or two local generators, each rated 2MW, at the load point. Consider the data in Table 7.10. The results for the present system with a load of 10MW and the two possible reinforcements with a load of 12MW are shown in Table 7.11.

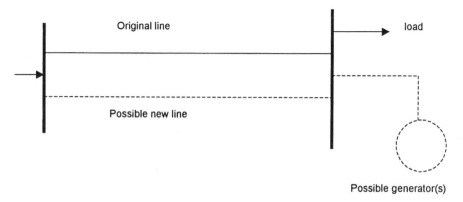

Figure 7.7   System to compare alternative reinforcements

Table 7.10   *Reliability data for system of Figure 7.6*

Component	Failure rate, f/year	Repair time, h
Feeders	0.5	24
Generators	2.0	96

Table 7.11   *Reliability results for system of Figure 7.6*

Failure	Failure rate (f/year)	Average down time (h)	Annual outage time (h/year)	Load disconnected (kW)	Energy not supplied (kWh/year)
Existing system					
Line	0.5	24	12	10	120
New system with 2 lines					
Line 1	0.5	24	12	2	24
Line 2	0.5	24	12	2	24
Total	1.0	24	24	2	48
New system with 1 line and 1 generator					
Line	0.5	24	12	10	120
Generator	2.0	96	192	2	384
Total	2.5	81.6	204	2.47	504
New system with 1 line and 2 generators					
Line	0.5	24	12	10	120
2 generators	0.084	48	4.03	2	8.1
Total	0.584	27.4	16.03	7.99	128.1

It can be seen from Table 7.11 that, before the load growth, the failure rate is 0.5 f/year, the annual unavailability is 12h/year and the expected energy not supplied (EENS) is 120kWh/year. After load growth and using two lines doubles the expected number of failures and annual unavailability but reduces the EENS to approximately one-third of the value for the existing system. Using one generator instead of a line increases all these indices quite severely (2.5f/year, 202h/year and 504kWh/year). However, by using two generators, the indices are approximately the same as for the existing system. Clearly these results are illustrative only and indicate the type of results that can be obtained. In reality, the actual indices will depend greatly on the data pertaining to the real system.

7.8 Generation reliability modelling

7.8.1 Modelling assumptions and considerations

The examples and case studies described in Section 7.7 do not reflect the stochastic nature of all factors affecting the behaviour of embedded generation, and the models must be extended. These are described in this and the following sections, and are based on the techniques described in References 35–37, where more details are provided.

As discussed in Section 7.5.2, energy sources can be classified into one of two types. The most difficult to deal with are those having an output dependent on unpredictable external energy sources such as wind and sun. The others can generally be modelled using conventional approaches. Therefore this chapter concentrates on the modelling requirements of these more complex and intermittent energy sources. It also considers specifically the case of wind energy sources because of their present popularity. However, the concepts are equally applicable to other intermittent energy sources such as solar since, in principle, a phrase associated with wind energy has an equivalent for other forms, e.g. wind turbine and solar panels, wind farm and solar plants, wind speed and solar radiation, etc.

Consequently this chapter concentrates on describing models for reliability studies that can take into account the stochastic nature of wind (and other sources), the failure and repair processes of the wind turbines, their output curves, wind speed spatial correlation and wake effects using reliability analysis based on analytical techniques. This model can be easily integrated with available models of the distribution network and can also be used for standard generation level (HLI) studies allowing and including a full reliability assessment of the system.

The discussion focuses on the wind farm perspective looking from inside the wind farm into the network. The results are integrated into reliability assessment techniques. Therefore, appropriate reliability results are evaluated at the wind farm boundary and also at any network/system boundary beyond. These results form the primary set of indices needed by all the parties involved to identify the effect of wind generation on their activities. Security domain aspects such as voltage dips, voltage flicker, transient stability and induction engine self-excitation are not considered.

7.8.2 Concepts of modelling

The total output of a generation station or overall plant (e.g. wind farm) is obtained by appropriately aggregating the outputs of each individual generator. The conventional approach to aggregate these outputs using probabilistic analytical techniques is the convolution of their individual capacity outage probability tables [12, 13]. This approach is applicable to non-intermittent sources such as gas, diesel and even CHP, but is not valid for wind generators or solar plant. The problem is that the outputs of units such as wind turbines are dependent on a common source, e.g. the wind. Therefore there is statistical dependence between the different generation output states, whereas independence is an underlying assumption in the convolution of capacity outage probability tables. Consequently it is not possible to calculate a generation output capacity table by convolving the output table of each wind turbine. The wind farm has to be considered in its entirety.

Determination of the statistical output characteristics of a wind farm for reliability analysis requires the simultaneous consideration of all wind turbines. Therefore a new wind farm generation model is needed which represents the stochastic characteristics of all processes involved. This can be accomplished by merging a wind (energy source) model and a wind turbine (generation) model.

7.8.3 Energy source model

Wind speed is a continuous physical phenomenon that evolves randomly in time and space. A random variable can be associated with each value of time. A stochastic process is considered to be a model of a system which develops randomly in time according to probabilistic laws [40]. Thus, wind speed is a stochastic process with a continuous state space (wind speed values) and continuous parameter space (time). This type of process can be modelled approximately as a discrete state space and continuous parameter state process.

Many studies have reported statistical tests on wind speeds using different probability distributions (Weibull, Rayleigh, χ^2, . . .). It is generally accepted that the Weibull distribution adequately represents the wind speed probability distribution [41–43] for most sampling times. This is all that is required to estimate the wind farms' expected output. For reliability analysis, however, the probability and frequency distributions of the wind speed stochastic process are also required.

The model used [35–37] to represent wind speed is a birth and death Markov chain [44, 45] with a finite number of states. The following assumptions are made in this model:

(a) The annual wind speed values in a period (e.g. annual, seasonal, etc.) are represented by a set of N wind speed levels or wind speed states.
(b) The wind speed model is statistically stationary.
(c) The distribution of residence times in a given state of the birth and death process is exponential.
(d) The probability of a transition from a given wind speed state to

another state is directly proportional to the long-term average (i.e. quiescent or steady state) probability of existence of the new state.
(e) Transitions between wind speed states occur independently of transitions between wind turbine states.
(f) From a given wind speed state, only the case of transitions to immediately adjacent (lower and higher speed) states are considered.

The wind speed range of interest is divided into a finite number of states that do not necessarily have to be equally spaced. This is an advantage because of the non-linear characteristic of the turbine output curve described in Section 7.8.4. The parameters of the wind model are calculated from a wind speed record. Wind speed is usually sampled at regular intervals. Available data are normally values of wind speed observations averaged over 10 min periods. The data that need to be extracted from the sample record to calculate the model parameters are the number of transitions from state i to $i \pm 1$, and the duration of the residence time in a state before going to a different state.

The non-continuous sampling process and the average values obtained for each observation mean that some transitions from recorded data could occur between non-adjacent wind speed states. This is an important consideration because wind speeds do not increase or decrease instantaneously, but change continuously, albeit over very short periods of time. Therefore the speeds will encounter all intermediate states even if not monitored. If this is the case, the duration of the intermediate states is estimated by a linear proportion of the sampling time.

If the wind speed duration follows an exponential distribution, then the transition rate between any two states can be calculated as

$$\lambda_{ij} = \frac{N_{ij}}{D_{ij}} \tag{7.1}$$

where N_{ij} is the number of transitions from state i to state j, and D_{ij} is the duration of state i before going to state j. To make a good parameter estimation it is necessary that the number of interstate transitions is as large as possible.

If the wind speed duration follows an unknown distribution then the parameters are calculated from the principle of frequency balance between any two states. The transition rates are calculated as

$$\lambda_{ij} = \frac{F_{ij}}{P_i} \tag{7.2}$$

where P_i is the state probability given by

$$P_i = \sum_{j=1}^{N} D_{ij} \bigg/ \sum_{k=1}^{N} \sum_{j=1}^{N} D_{kj} \tag{7.3}$$

F_{ij} is the frequency of transitions between state i and state j and N is the total number of states.

The main advantage of this model is that it represents exactly the real characteristics of the wind speed found at a particular site with a representation suitable for conventional reliability analysis. Hence, regardless of the sampling time, if the wind speed range to be considered is discretised into a finite number of states, it is always possible to find a finite birth and death process with the same number of states that reproduces exactly the probability and frequency distributions of the sample.

7.8.4 Generation model

The wind turbine model defines the output characteristics of the wind turbine as well as its failure and repair processes. The wind turbine output curve determines the power output of the machine for corresponding wind speed levels. The wind turbines can be divided into two major types according to the type of output control: pitch or stall regulated wind turbines (Figure 7.8). Both can be represented by the proposed wind turbine model [35–37].

The failure and repair rates of the individual wind turbines should be taken into account to calculate the expected wind farm output. Failures under extreme wind speeds are normally of a catastrophic nature such as cracking of the tower and blades. These are lengthy and costly failures with a low probability but whose expected frequency should be carefully considered. Therefore two major wind speed ranges can be distinguished:

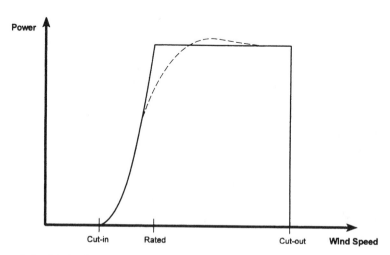

Figure 7.8   *Typical output curves for wind turbines*
——— pitch control, – – – – stall control

a range within the designed for wind speeds and another above it comprising extreme wind speeds. Consequently, different sets of failure and repair rates are considered within each range.

The wind turbine is modelled as a binary state component, i.e. the machine is considered fully capable of giving any output within its power curve limits (up) or it is out of service (down). The failures of different wind turbines are considered independent. Therefore the resulting state transition diagram for a given number of turbines is equal to that obtained when a system with the same number of independent components is considered. Wind generators are relatively simple machines and so the maintenance time is much less than that of conventional units and therefore is not included in the models.

7.8.5 Generation plant model

The wind farm model comprises the wind model and the wind turbine model. For each of the wind states considered, the associated wind turbines' failure and repair processes are represented [35–37]. States with the same wind turbine contingency are connected by the wind model transition rates. The resulting diagram can be viewed as a set of layers (wind turbines states) connected by the wind model transition rates (Figure 7.9 for two wind states, I and II). For each wind farm state, the number of wind turbines in service is known, as is the corresponding wind speed

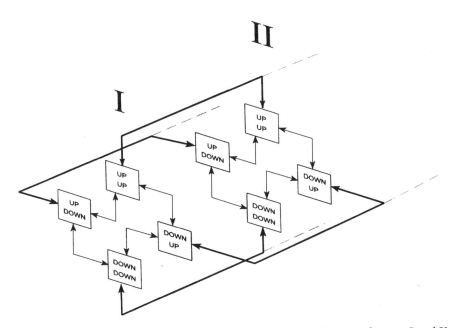

Figure 7.9   Wind farm state diagram for two turbines and two wind states I and II

level. Therefore it is straightforward to determine the output of each wind turbine by using the output curves. The total wind farm output for each specified wind farm state is obtained by adding the output of all wind turbines while in that state. This solves the problem of statistical dependence to calculate the wind farm output.

In general the transition rates for all states in each wind speed layer could be different, although it seems realistic to consider only two sets of values, one for normal and the other for extreme wind speeds, as stated before. The impact of failures in the elements connecting the different wind turbines inside the wind farm is neglected.

The main advantages of this model are:

- The addition of more wind state levels to increase the precision of power output and energy calculations is simple.
- More turbines can easily be added to the wind farm.
- It is always theoretically possible to have different sets of transition rates for each wind state.
- States with different but non-zero outputs are not directly connected, so that the cumulation of identical output capacity states is simple.
- The model can be combined with existing reliability models of the distribution network to evaluate the overall impact of the wind farm.

Several models have been studied [35–37], some with extensive complex configurations. However, the model shown in Figure 7.10 proved to be the best compromise between accuracy and computational effort. This model does not truly reflect correct behaviour with respect to state 8, which can be entered in two ways. However, extensive assessments showed that this model tends to give slightly more pessimistic results due to the prevailing effect of transitions from minor damage states to severe damage states over the opposite direction transitions.

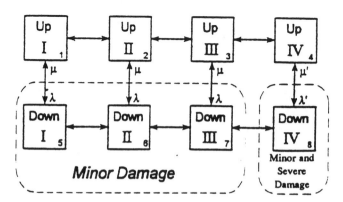

Figure 7.10 Wind farm model of one turbine and four wind levels
λ = failure rates, μ = repair rates

If a wind turbine is exposed to the turbulent air wake of another turbine, the amount of power that can be extracted is less than that if it was exposed to free air flow. This effect is minimised for a carefully laid out wind farm. However, if the wind has significant variations in direction, this effect might be significant for those occasions when the hubs turn to face particular wind directions. To consider this effect, a correction factor in the power generated is applied to those wind turbines more likely to experience this effect, the value being very dependent on the design of the wind farm. The possibility that a turbine affected by waking can extract more power when the waking turbine is out of service is neglected for the same reasons given above regarding maintenance and repair times.

Different types of wind turbines having different output curves and/or failure and repair characteristics are treated as different components in the layers. In this way it is simple to add the effects of spatial correlation or wake effects as new turbines (components) with different output curves and/or failure and repair parameters.

7.8.6 *Solution of the plant model*

Solution of the wind farm model involves calculating the probability and frequency of each state in the system. The limiting state probabilities can be found from the stochastic transitional probability (STP) matrix [19] of the system. Two problems arise in solving the proposed wind farm model as the number of wind turbines and/or wind states increases. These are: (i) reduction of the number of states resulting from the wind farm model; (ii) construction of the STP matrix representing the connections between the different states.

A detailed discussion and description of the techniques used to solve this Markov model are described in References 35–37. This solution provides the information needed to calculate the expected output of the wind farm which may be the objective of the assessment. However it also provides the input to the overall reliability assessment of the distribution system as described in the following sections.

7.9 Network reliability model

The following discussion also focuses on wind energy sources as a continuation of Section 7.8. However, all the principles apply equally to all generation sources that have been modelled in a way that creates an appropriate generation capacity outage probability table. This may have been obtained using conventional techniques assuming independence (gas, oil, etc) or using the approach described in Section 7.8 for intermittent sources (wind, solar, etc).

As discussed in Section 7.4.1, faults in the distribution network are frequently the most important cause of unavailability to customers accounting often for more than 80% of the time without service. Most wind farms are connected to distribution networks because their sizes do not justify investment in a high voltage connection to the grid. As the best wind resources are normally found in remote areas, they are usually connected to weak (radial) distribution networks. Therefore the combination of these two factors may pose limitations to the maximum expected output of the wind plant. Questions then arise about the effects of network reinforcement and of different operational policies on the ability to extract energy from the wind plant, on the benefit derived by customers, on the impact on the next voltage level and normal supply of energy, etc. Answers to these questions can only be objectively given from a reliability assessment of the distribution network containing the wind generation.

The conventional approach [12] to evaluating distribution system reliability is a load point orientated one based on minimal cut set theory, i.e. a load point is selected, relevant outages for that load point identified and load point indices evaluated. This is the principle used to evaluate column-by-column the load point indices given in Table 7.5, for example. This approach is implemented in two stages: total loss of continuity (TLOC – does not consider overloadings), valid for radial networks, and partial loss of continuity (PLOC – requires load flow subroutines to evaluate overloadings), for meshed networks.

Instead of using the load-point driven approach, a method similar to that used for assessing transmission systems can be used [36, 37, 46]. This is an event-driven approach which proves more appropriate for distribution systems containing embedded generation. Instead of selecting a load point and studying all events that could lead to interruption of supply at that load point, a contingency event is considered and all the load points that are affected by it are deduced simultaneously. In principle, it corresponds to deducing the results given, for example, in Table 7.5 row-by-row rather than column-by-column. Each contingency is a particular system state which, when embedded generation exists in a distribution system, is made up from a wind generation state, a load state and a network state.

The generation states are given by the output of the wind farm model described in Section 7.8.6 or obtained from conventional convolution for non-intermittent sources. The load states can be deduced from load levels sequentially predicted for the system being analysed or from any other suitable model. In all cases, the probability, frequency and duration of load states can be evaluated.

The network model defines state events and failure modes similar to those used in conventional reliability models of distribution systems. These include failure of busbars, lines, transformers, breakers and fuses, the isolation of a failed component, the transfer of load through normally open points, the assessment of both radial and meshed systems, and the consideration of TLOC and PLOC.

The procedure to identify the effect of network contingencies is based on graph theory used in general network flow problems [47]. This uses arcs and nodes to determine connectivity. A full description is given in References 36, 37 and 46. When arcs are opened either by the operation of the protection system or by the isolation actions for the repair of failed components, a forest of trees can be created in the network. These trees are processed to give the number of islands created in the system due to the contingency being considered. It also gives the load points belonging to each island and where the generation sources are situated. With this information, the corresponding reliability calculations can be performed.

7.10 Reliability and production indices

To express the results of the reliability and production analysis an adequate set of indices is required. With careful interpretation these can be used to quantitatively compare different systems and to measure the benefit of alternative designs, strategies and impact of embedded generation.

7.10.1 Capacity credit

One measure that has become associated with embedded generation is the term 'capacity credit'. This is said to be the effective capacity for which the generation source can be credited and is evaluated by dividing the energy generated in a period by the number of hours in that period. It therefore gives the average capacity that the system sees measured over the period considered. Although this may seem to provide a useful measure, it is only useful to a generator as an indicator of overall contribution but is flawed as a useful reliability index for either customers or energy suppliers, particularly in connection with intermittent sources such as wind. To illustrate one weakness, consider a capacity credit of say 25%, which may be considered typical for a wind farm. This implies that, on average, it supplies 25% of the installed capacity continuously, whereas in real life it could supply either full capacity (four times its capacity credit) for 25% of the time and nothing for 75% of the time. This may be much less of a problem for sources not dependent on the environment and which can be scheduled and controlled, in which case the value of capacity credit becomes very large.

It can also be concluded that, because of their intermittent nature, such sources cannot replace conventional generation and are really energy-replacement rather than capacity-replacement sources [33]. As wind power plants with higher installed capacities are incorporated into existing power systems, the problem may become even more emphasised. The diversity created by using other forms of embedded generation could lessen the effect of no wind. However, it follows that it is becoming increasingly important to study the reliability of these generation systems and assess the effects that they will have on the entire system and its reliability. To achieve this, additional and more meaningful indices and measures are therefore required.

7.10.2 Reliability indices

There are many reliability indices that can be calculated and used in assessing distribution systems. These include the load point indices and system indices [12, 13] as already discussed in Section 7.4.2. The most appropriate depends on the perspective viewed by each party involved. Energy-based indices may be the most appropriate from a generator's perspective because these permit assessment of income derived, but interruption-based indices may be more appropriate from an end customer's perspective because these determine their ability or inability to use the supply. It is therefore important to be able to evaluate a range of indices and choose the most relevant depending on the circumstances and objective of any particular decision process.

7.10.3 Production indices

Generation indices have been defined to reflect the specific aspects affecting the performance of embedded generation such as wind. These include, in the case of wind [35–37, 46] (other sources have direct equivalents):

- installed wind power (IWP) – sum of the rated power of all the wind generators in the wind farm
- installed wind energy (IWE) – installed wind power multiplied by a year, this is the energy that could be extracted if the units could be operated continuously (IWP × 8760)
- expected available wind energy (EAWE) – expected amount of energy that would be generated in a year if there were no wind turbine generator (WTG) outages
- expected generated wind energy (EGWE) – expected maximum amount of energy that would be generated in a year by the real WTGs considering their outage rates and the real wind to which they are exposed
- wind generation availability factor WGAF: EGWE/IWE
- wind generation utilisation factor WGUF: EWEU/EGWE
- capacity factor (≡ capacity credit): WGAF × WGUF.

7.11 Study cases

To demonstrate the application of the preceding techniques and to illustrate the types of results that can be obtained, a study extracted from References 36, 37 and 46 is described in this section. The network used is shown in Figure 7.11. It is a 33/11 kV meshed system connected to the 132 kV network by two parallel transformers. The total load is 82.55 MW. The inclusion of a wind farm made up of 15×1 MW machines and connected to bus 9 as shown was considered. Typical wind, wind turbine, load profile and reliability data were used. The reliability indices for the load buses, including interruption and energy indices, are shown in Table 7.12.

The following observations can be made. The load point reliability indices for the split busbars 2 and 5 are the same. However, the indices for the split busbar 3 are different. This is because the lines between busbars 2b and 3b (lines 2a–3a and 2b–3b) and the line between busbars 2a and 3a have different reliability characteristics. The position of the breakers

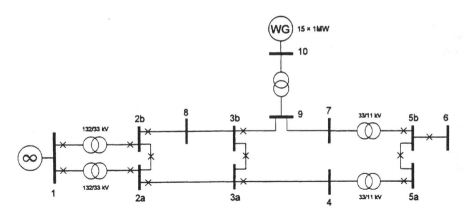

Figure 7.11   Typical 33/11kV system with one wind farm

Table 7.12   Load point indices for system of Figure 7.11

Load point	Load (MW)	λ (f/year)	U (h/year)	r (h)	ENS (MWh)
2a	28.75	0.0000	0.0019	82.550	0.034
2b	28.75	0.0000	0.0019	82.550	0.034
3a	7.85	0.7585	12.726	16.802	59.98
3b	7.85	0.8400	14.096	16.808	66.44
5a	4.50	0.0110	0.1181	10.784	0.319
5b	4.50	0.0110	0.1181	10.784	0.319
6	0.35	0.6160	3.1420	5.1029	0.660

between the busbars means that the failure of these lines affects busbars 3a and 3b independently and thus explains the difference in the reliability indices.

The system reliability indices are shown in Table 7.13, and the wind generation indices are shown in Table 7.14. The EEWE is calculated at the system boundary with the next voltage level (bus 1). In this case all the energy generated by the wind farm is consumed within the system being assessed. This is because the minimum load level in the network exceeds the maximum capacity of the wind farm: otherwise some energy would be available for export. The wind generation indices show that the effect of failures in the interfacing network introduces restrictions to the wind farm output (WGUF = 0.9968). The effect, however, is quite small.

The energy statistics for this system are shown in Table 7.15. The main effect of the introduction of the wind farm is a reduction in power and energy losses in the system. Initially the expected energy losses were 4232 MWh, about 0.97% of the total energy consumed in the system. When

Table 7.13   System reliability indices for system of Figure 7.11

Index	Value
SAIFI	0.156 int/customer.year
SAIDI	2.58 h/customer.year
CAIDI	16.54 h/customer int
ASAI	0.999 706
ASUI	0.000 294
ENS	127.8 MWh/year
AENS	0.015 MWh/customer.year

Table 7.14   Wind generation indices for system of Figure 7.11

Index	Value
IWP	15.0 MW
IWE	131 400 MWh/year
EAWE	35 333 MWh/year
EGWE	34 642 MWh/year
EWEU	34 530 MWh/year
EEWE	0 MWh/year
Availability	0.98042
WGAF	0.2636
WGUF	0.9968
WGAF × WGUF	0.2628

Table 7.15   Energy balances with and without wind farm for system of Figure 7.11

Energy (MWh)	Without wind farm	With wind farm
Consumed	434 045	434 045
Net imported	438 278	402 873
		(−8.08%)
Exported	0.0	0.0
Losses	4 232.4	3 357.5
		(−20.68%)
Wind farm exported	0.0	34 530

the wind farm is included, the energy losses are reduced by 20.68%. Thus, the immediate effect of the wind farm is an instant increase in the efficiency of the system operation. Moreover, in this case the net imported energy from the 132 kV system was 101% of the total energy consumed. However, this figure is reduced by 8.08% to 92.8% of the energy consumed. Thus, a reduction in energy purchase from the 132kV system of more than 8% can be expected after the introduction of wind generation.

7.12 Conclusions

Many distribution systems are still designed according to deterministic standards. These views are changing quite significantly and there is now a positive awareness of the need to assess system design alternatives in a probabilistic sense. This awareness is increasing following the restructuring of the electricity supply industry since it is now necessary for planning and operation decisions to be transparent and equitable to all parties involved. This is particularly so in distribution systems in which many interested parties have conflicting interests, including the actual end customers, the network owners and operators, the energy suppliers, the owners and operators of the transmission network, the conventional generators, as well as new entrants concerned with embedded generation. The ability to compare alternatives objectively is becoming a necessity, and this can only be achieved quantitatively using probabilistic assessment approaches. This chapter has addressed the issues involved in such assessments and described various models and techniques which reflect the stochastic nature of the energy sources, the generating plant and customer demands, and which can be used to perform such assessments.

In addition, there is also a rapidly growing appreciation, inside and outside the industry, of the need to account for customers' expectations and their assessment of the worth of supply. Since the latter cannot be objectively assessed without quantitative reliability measures, the inter-relation between the two aspects of reliability and worth of supply is also expected to become of significant importance in the very near future. This will lead to extended use of cost–benefit analyses.

The detailed results quoted in this chapter show that the approaches described provide valuable information relating to the energy-exporting capability of an embedded generating plant (information of benefit to the plant owner and operator), the energy consumed by the end cus-tomers in the distribution system (of benefit to the distribution company, energy suppliers and consumers), and the energy available for export to other systems and the grid (of benefit to distribution and grid com-panies). The results also show that the reliability indices of customer load points do not change significantly when embedded generation is not permitted to operate as stand-alone units. However, one particular tech-nical benefit is that there can be a significant reduction in system losses which is of direct benefit to the distribution company hosting the embedded generator. This, together with the energy-replacement feature of embedded generation, forms some of the most significant advantages of these sources of energy.

7.13 References

1 ELECTRICITY COUNCIL: 'Security of supply'. Engineering Recom-mendation P2/5, October 1978
2 OFFICE OF ELECTRICITY REGULATION: 'Report on customer ser-vices 1997/98'. OFFER, 1998
3 BILLINTON, R.: 'Bibliography on the application of probability methods in power system reliability evaluation', *IEEE Transactions on Power Appar-atus and Systems*, 1972, **PAS-91**, pp. 649–660
4 IEEE SUBCOMMITTEE: 'Bibliography on the application of probability methods in power system reliability evaluation, 1971–1977', *IEEE Transac-tions on Power Apparatus and Systems*, 1978, **PAS-97**, pp. 2235–2242
5 ALLAN, R.N., BILLINTON, R., and LEE, S.H.: 'Bibliography on the application of probability methods in power system reliability evaluation, 1977–1982', *IEEE Transactions on Power Apparatus and Systems*, 1984, **PAS-103**, pp. 275–282
6 ALLAN, R.N., BILLINTON, R., SHAHIDHPOUR, S.M., and SINGH, C.: 'Bibliography on the application of probability methods in power system reliability evaluation, 1982–1987', *IEEE Transactions on Power Systems*, 1988, **PWRS-3**, pp. 1555–1564
7 ALLAN, R.N., BILLINTON, R., BRIEPOHL, A.M., and GRIGG, C.H.: 'Bibliography on the application of probability methods in power system reliability evaluation, 1987–1991'. IEEE Winter Power Meeting, Columbus, February 1993, Paper 93 WM 166–9 PWRS
8 ALLAN, R.N., BILLINTON, R., BRIEPOHL, A.M., and GRIGG, C.H.: 'Bibliography on the application of probability methods in power system

reliability evaluation, 1991–1996'. IEEE Winter Power Meeting, 1998, Paper PE-201-PWRS-0–2–1998

9 BILLINTON, R., ALLAN, R.N., and SALVADERI, L. (Eds.): 'Applied reliability assessment in electric power systems' (IEEE Press, New York, 1991)

10 ALLAN, R.N., and BILLINTON, R.: 'Probabilistic methods applied to electric power systems. An important question is – Are they worth it?', *Power Engineering Journal*, 1992, **6**, pp. 121–129

11 BILLINTON, R., and ALLAN, R.N.: 'Power system reliability in perspective', *Electronics and Power*, 1984, **30**, pp. 231–236

12 BILLINTON, R., and ALLAN, R.N.: 'Reliability evaluation of power systems' (Plenum Publishing, New York, 1984, 2nd edn.)

13 BILLINTON, R., and ALLAN, R.N.: 'Reliability assessment of large electric power systems' (Kluwer Academic Publishers, Boston, 1988)

14 CIGRE WORKING GROUP 38.03: 'Power system reliability analysis – Application Guide' (CIGRE Publications, Paris, 1988)

15 CIGRE WORKING GROUP 38.03: 'Power system reliability evaluation: Vol.2, Composite power system reliability evaluation' (CIGRE Publications, Paris, 1992)

16 CIGRE TF 38.03.11.: 'Methods and techniques for reliability assessment of interconnected systems'. CIGRE Technical Brochure 129, 1998

17 CIGRE TF 38.03.12.: 'Power system security assessment: A position paper', *Electra*, December 1997, (175), pp. 49–78

18 OFFICE OF ELECTRICITY REGULATION: 'Report on distribution and transmission system performance 1997/98'. OFFER, 1998

19 BILLINTON, R., and ALLAN, R.N.: 'Reliability evaluation of engineering systems: Concepts and techniques' (Plenum Publishing, New York, 1992, 2nd edn.)

20 ALLAN, R.N., and M. DA GUIA DA SILVA.: 'Evaluation of reliability indices and outage costs in distribution systems', *IEEE Transactions on Power Systems*, 1995, **PWRS-10**, Paper 94 SM 576–9-PWRS

21 ALLAN, R.N., DIALYNAS, E.N., and HOMER, I.R.: 'Modelling and evaluating the reliability of distribution systems', *IEEE Transactions on Power Apparatus and Systems*, 1979, **PAS-98**, pp. 2181–2189

22 KARIUKI, K.K., and ALLAN, R.N.: 'Assessment of customer outage costs due to service interruptions: residential sector', *IEE Proceedings – Gener. Transm. Distrib.*, 1996, **143**, pp. 163–170

23 KARIUKI, K.K., and ALLAN, R.N.: 'Evaluation of reliability worth and value of lost load', *IEE Proceedings – Gener. Transm. Distrib.*, 1996, **143**, pp. 171–180

24 KARIUKI, K.K., and ALLAN, R.N.: 'Application of customer outage costs in system planning, design and operation', *IEE Proceedings – Gener. Transm. Distrib.*, 1996, **143**, pp. 305–312

25 KARIUKI, K.K., and ALLAN, R.N.: 'Factors affecting customer outage costs due to electric service interruptions', *IEE Proceedings – Gener. Transm. Distrib.*, 1996, **143**, pp. 521–528

26 WACKER, G., and BILLINTON, R.: 'Customer cost of electric service interruptions', *Proc. IEEE*, 1989, **77**, pp. 919–930

27 BILLINTON, R., WACKER, G., and WOJCZYNSKI, E.: 'Customer

damage resulting from electric service interruptions'. Report on Canadian Electric Association R&D Project 907 U 131, 1982

28   DESHMUKH, R.G., and RAMAKUMAR, R.: 'Reliability analysis of combined wind-electric and conventional generation systems', *Solar Energy*, 1982, **28**, (4), pp. 345–352

29   GIORSETTO, P., and UTSUROGI, K.F.: 'Development of a new procedure for reliability modeling of wind turbine generators', *IEEE Transactions on Power Apparatus and Systems*, 1983, **PAS-102**, (1), pp. 134–143

30   WANG, X., DAI, H., and THOMAS, R.J.: 'Reliability modeling of large wind farms and associated electric utility interface systems', *IEEE Transactions on Power Apparatus and Systems*, 1984, **PAS-103**, (3), pp. 569–575

31   SINGH, C., and LAGO-GONZALEZ, A.: 'Reliability modelling of generation systems including unconventional energy sources', *IEEE Transactions on Power Apparatus and Systems*, 1985, **PAS-104**, (5), pp. 1049–1056

32   SINGH, C., and KIM, Y.: 'An efficient technique for reliability analysis of power systems including time dependent sources', *IEEE Transactions on Power Systems*, 1988, **3**, (3), pp. 1090–1096

33   ALLAN, R.N., and CORREDOR AVELLA, P.: 'Reliability and economic assessment of generating systems containing wind energy sources', *IEE Proc. C*, 1985, **132**, (1), pp. 8–13

34   BILLINTON, R., and CHOWDHURY, A.A.: 'Incorporation of wind energy conversion systems in conventional generating capacity adequacy assessment', *IEE Proc. C*, 1992, **139**, (1), pp. 47–56

35   CASTRO SAYAS, F., and ALLAN, R.N.: 'Generation availability assessment of wind farms', *IEE Proceedings – Gener. Transm. Distrib.*, 1996, **143**, (5), pp. 507–518

36   CASTRO SAYAS, F.: 'Reliability assessment of distribution networks containing embedded wind generation'. PhD thesis, UMIST, 1996

37   ALLAN, R.N., and CASTRO SAYAS, F.: 'Reliability modelling and evaluation of distribution systems containing wind farms'. Proceedings of International Conference on *Probabilistic Methods Applied to Power Systems* (PMAPS'97), Vancouver, Canada, 1997, pp. 181–186

38   DEVRIOU, A., SCHNEIDER, M., DANIELS, G., and TZCHOPPE, J., 'Influence of dispersed generation on networks reliability'. CIGRE Symposium on *Impact of DSM, IRP and distributed generation on power systems*, Neptun, Romania, 1997, Paper 300–03

39   HATZIARGYRIOU, N.D., PAPADOPOULOS, M.P., SCUTARIU, M., and EREMIA, M., 'Probabilistic assessment of the impact of dispersed wind generation on power system performance'. CIGRE Symposium on *Impact of DSM, IRP and distributed generation on power systems*, Neptun, Romania, 1997, Paper 300–15

40   PAPOULIS, A.: 'Probability, random variables and stochastic processes' (McGraw-Hill, New York, 1991, 3rd edn.)

41   FRERIS, L.L.: 'Wind energy conversion systems' (Prentice Hall, London, 1990)

42   EGGLESTON, D.M., and STODDARD, F.S.: 'Wind turbine engineering design' (Van Nostrand Reinhold, New York, 1987)

43   JOHNSON, G.L.: 'Wind energy systems' (Prentice Hall, Englewood Cliffs, NJ, 1985)

44   ANDERSON, W.J.: 'Continuous-time Markov chains' (Springer Verlag, New York, 1991)
45   CHUNG, L.L.: 'Markov chains with stationary transition probabilities' (Springer Verlag, New York, 1967)
46   'Technical and economic effects of connecting increasing numbers of large wind turbines to weak public utility networks'. Report JOU2-CT92-0095. CEC DG XII
47   JENSEN, P.A., and WESLEY BARNES, J.: 'Network flow programming' (Krieger Publishing Co, USA, 1987)

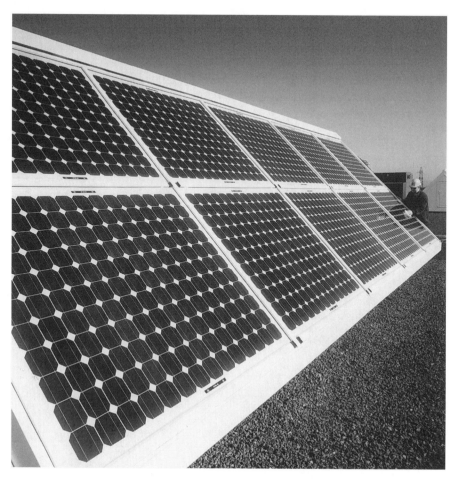

BP Solar (high efficiency) photovoltaic panels integrated into the roof of Ford Motor Co., Bridgend. This is part of a 97 kWp array with an average energy generation of 77.6 MWh per annum. Multiple 3.7 kW force-commutated inverters are used for connection to the AC system of the factory.

Photo: Chris Ridley

Chapter 8

Economics of embedded generation

8.1 Introduction

As discussed in Chapter 1, the presence of embedded generation (EG) in distribution systems alters radically the way these networks should be viewed from both technical and commercial vantage points. This is because EG effectively turns distribution networks from passive networks with unidirectional power flows from higher to lower voltage levels into active networks with multidirectional power flows. This change challenges the validity of traditional distribution network planning, operation and commercial practices, in which distribution networks are treated as passive. For the value of EG to be appropriately recognised and hence its development encouraged, it is essential to factor the active nature of distribution networks with EG into all commercial as well as technical operation and planning activities.

An essential condition for competition to develop is open access, on a non-discriminatory basis, to transmission and distribution networks. The central issue in the concept of open access is setting an adequate price for transmission and distribution services as this affects future siting of generators and loads, network operating costs and service quality delivered by the networks, and strongly influences further network development. Under such a scenario, there is ever growing pressure for all components of costs to be clearly identified and assigned efficiently and equitably to all parties avoiding temporal or spatial cross-subsidies.

To facilitate fair competition between various generators, central and embedded, setting of appropriate connection and use of transmission and distribution system tariffs, as well as an equitable loss allocation policy, is essential. Due to its location, embedded generation not only acts as another source of electricity but can potentially substitute for transmission and high-voltage distribution facilities as well as reduce losses in those networks. This is often used to argue that a kWh produced by EG has a higher 'value' than a kWh generated at transmission level by a conventional central generating plant. As embedded generation directly competes with central generation, it is essential to have a consistent commercial framework for pricing of network services to establish fair competition among generators.

In the context of the economic interaction between embedded generation and power networks, this chapter discusses the impact of EG on distribution network operating and capital costs. Particular emphasis is placed on the relationship between network pricing practices and the economic impact of embedded generation on power networks. The appropriateness of the present network pricing arrangements is discussed as well as possible future developments.

8.2 Connection costs and charges

8.2.1 Concept

To recover the investments necessary to enable consumers and generators to access the electricity marketplace, the notion of connection and use of system charges is used. Under the UK Electricity Act, Regional Electricity Companies (RECs) have a duty to provide a supply of electricity as required. In meeting such a request, the Companies may set connection charges at a level which enables them to recover the costs incurred in carrying out any works, the extension or reinforcement of the distribution system, including a reasonable rate of return on the capital represented by the costs. From the EG perspective, two questions related to the policy of connecting EG are of considerable importance:

(a) the voltage level to which generation should be connected, as this has a major impact on the overall profitability of generation projects
(b) the question of whether a connection policy is based on 'shallow' or 'deep' charges (shallow charges reflect only costs exclusively associated with making the new connection, while deep charges also include the additional costs which are indirectly associated with the reinforcement of the system).

These two issues are discussed in turn as follows.

8.2.2 Voltage level related connection cost

The overall connection costs may considerably alter the cost base of an embedded generator and are primarily driven by the voltage level to which the generator is connected. Generally, but not exclusively, the higher the voltage level the larger the connection cost. To secure the viability of a generation project, developers and operators of EG would prefer to be connected at the lowest possible voltage level. On the other hand, as indicated in Chapter 3, the higher the connecting voltage the lower the impact that embedded generation has on the performance of the local network, particularly in terms of steady-state voltage profile and power quality. Therefore, the network operators generally prefer connecting EG to higher voltage levels. These two conflicting objectives need to be balanced appropriately, and may require an in-depth technical and economic analysis of the alternative connection designs. A study published in Reference 1 illustrates the importance of acquiring a thorough understanding of a number of power system issues if the most cost-effective connection is to be realised. In this respect, the voltage rise effect which a generator connected to a weak network could produce is particularly critical. It is important to stress that in the majority of European countries the accepted steady-state voltage variations are much stricter than the European Norm, EN 50160 [2].

To which voltage level an embedded generator can be connected to the distribution network will largely depend on its size, but also on the layout of the local network, its parameters. The proximity of load may also be important, and it is therefore not possible to derive generalised rules. As listed in Chapter 1, generators up to 500kW could typically be connected to 415V networks, 5MW to 11kV networks, and up to 20MW to 33kV networks.

In practice, the level of penetration of EG depends largely on network connection rules. These will determine the amount of EG that can be connected to the local networks. For example, the design of connections is determined on a case-by-case basis in the UK following quality standards defined in engineering recommendations [3, 4]. In contrast to this practice, the connection rules in Germany [5] are derived from standards of network operation that assume typical medium voltage networks with average loads and typical lengths of lines. The latter approach brings some advantages in terms of administrative handling of the connections at the expense of the inability to capture the specifics of individual cases. Another interesting example is France, where generators between 10 and 40 MW may only be connected to networks above 50 kV, and plant greater than 40 MW must be connected at 225kV and above [6]. Clearly, these rules, though determining the voltage level at which generators should be connected to the networks, are crucial not only for the commercial success of individual generation projects, but also for the general level of penetration of embedded generation.

Reactive and/or active power control may be used as a means of controlling the network voltage profile and in particular the voltage rise effect. Currently, however, embedded generation does not actively participate in voltage regulation of distribution systems. This is because the present network tariff structure does not support this activity and an adequate commercial framework for the provision of voltage regulation services on a competitive basis has yet to be developed.

As an example, VAR management as a means of reducing voltage excursions in distribution networks is not encouraged by appropriate pricing mechanisms. Present reactive power pricing in some distribution networks is that generators that absorb reactive power are charged by the distribution company on the basis of the active power demand taken by the plant, not its generation. As active power input is much lower than the active power output, reactive power taken by the generator is seen as a reactive excess, and consequently, the generator is expected to pay the highest possible excess reactive charges. There are two different approaches to charging for reactive power/energy. The majority of UK RECs charge with respect to kVArh (reactive energy) in excess of 40–50% of the total unit consumption in that month. Typical values are given Table 8.1.

Some other RECs charge for maximum kVAr (reactive power) of demand in excess of the value obtained by multiplying the maximum kW of demand registered in any time during the month by 0.4. Typical values are given in Table 8.2. There are also RECs who base their distribution use-of-system charges on kVA demand, which discourages consumption of reactive power.

Absorbing reactive power can be very beneficial to controlling the voltage rise effect in weak overhead networks with embedded generation. Although this would normally lead to an increase in network losses, EG does not have the opportunity to balance the connection costs against cost of losses and make an appropriate choice. Clearly, the above tariff structure discourages generators from participating in voltage regulation. Conversely, synchronous generators are offered no incentive by the distribution company to provide reactive support and take part in voltage regulation. The philosophy behind the excess reactive charges has been derived for passive distribution networks and cannot easily be justified in the context of a distribution network with embedded generation.

Similarly, active power generation shedding could be used as a means of reducing the voltage rise effect. The generator may find it profitable to

Table 8.1   Reactivity energy based charges

Charges/voltage level	Low voltage	High voltage
Excess reactive charge (40–50%) p/kVArh	0.50–0.64	0.36

Table 8.2   Reactivity power based charges

Charges/voltage level	Low voltage	High voltage
Excess reactive charge (40%) £/kVAr month	1.63–3.11	1.30–1.52

shed some of its output for a limited period if allowed to connect to a lower voltage level. However, this option has not yet been offered to generators.

It follows from the above discussion that the inability of the present reactive power pricing concept to support provision of voltage regulation may unnecessarily force generators to connect to a higher voltage level, imposing significant connection costs. Further development of market mechanisms and pricing policies may lead to the development of a market for the provision of voltage regulation services in distribution networks and provide more choice for embedded generation to control its connection costs. This is an area that has only recently received attention. It is expected that it will develop in the near future and open up further possibilities for embedded generation to participate not only in the energy market but also in a market for the provision of ancillary services.

Appropriate tools, which would enable the establishment of an ancillary services market in distribution networks, are not currently available to network operators. This area is receiving increasing attention from both industry and academia.

8.2.3 Deep vs. shallow connection charges

Another issue that can significantly influence the profitability of a generation project is whether connection charges should reflect only costs exclusively associated with making the new connection or also include the additional costs which are indirectly associated with any reinforcement of the system. In other words, should connection charges be based on so-called shallow or deep connection costs?

The costs associated with connecting an EG to the nearest point in the local distribution network are referred to as shallow connection charges. Clearly, a circuit between the new EG and the system is used only by the generator. The generator is therefore required to cover the cost of this connection through connection charges (which could be imposed over a period of time if the distribution company invests and owns the connection, or alternatively, as a one-off payment, in which case the generator is effectively the owner of the line).

It has been argued that the advantage of shallow connection charges lies in the simplicity of their definitions. Also, it is relatively straightforward to identify the cost related exclusively to connecting the generator to the nearest point in the network. On the other hand, the cost associated with a new connection might not be fully reflected in the connection charge made, as such a connection may require reinforcement of the system away from the connection itself. The majority of RECs charge new entrants for the cost of connection itself and the cost incurred for any upstream reinforcement. Following the Distribution

Price Control this is limited to one voltage level up from the point of connection.

Such a situation is presented in Figure 8.1. For the purpose of illustrating the concept of deep charging, it is assumed that the new connection requires the circuit breaker at the higher voltage level to be replaced, as the presence of the new generator increases the fault level. If the embedded generator is required to cover the cost of the replacement of the circuit breaker, this would be referred to as a deep connection charge.

Now the size of this circuit breaker is determined by the contribution of all generators, both central and embedded. The individual contribution of each generator to the size can be readily computed using conventional fault analysis tools, as indicated in Chapter 3. These contributions to the short-circuit current may then be used to allocate the cost of replacing the circuit breaker. This is indicated in Figure 8.2.

An argument such as 'there would not be a need to replace the breaker if the new generator did not appear and therefore the new entrant should be responsible for all costs' cannot be credibly used to require the new entrant to recover all system reinforcement costs. In accordance with conventional economic theory it can be argued that the distribution network owner should replace the circuit breaker, and recover its cost through charging all generators according to their respective contributions. This could be achieved by adjusting the use of system charges to all generators in the following price review period. The circuit breaker in question could be considered as a system related investment, and hence its cost should be recovered through the use of system charges rather than through the connection charges. Under such a scenario, large gen-

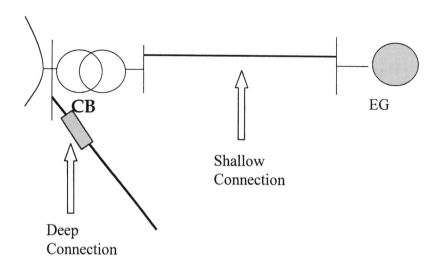

Figure 8.1   Illustration of shallow and deep connection

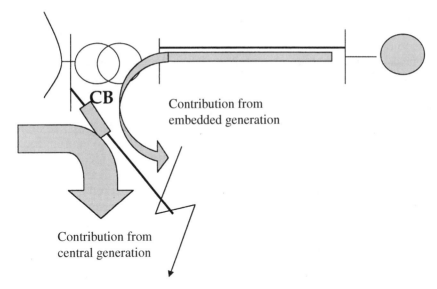

Figure 8.2   Contribution of central and embedded generators to fault level

erators connected to the transmission system are likely to be required to contribute most to these costs. It is, however, important to observe that, in accordance with the present practice for pricing of network services, central generators are not charged for the use of distribution networks, as if these networks were not required for the use of their output.

8.3 Distribution use of system charges and embedded generation

8.3.1 Current practice

The distribution network business is dominated by capital cost and operates in a near monopolistic or highly restricted competitive environment. To recover the costs involved in serving its users and to facilitate competition in supply and generation of electricity, tariffs are set for using distribution networks.

Currently used distribution use-of-system (DUoS) tariffs have been developed for customers who take power from the network rather than for customers who inject power into the network. For the evaluation of charges for use of distribution networks the majority of distribution companies use a model known as the Distribution Reinforcement Model. This model is employed to evaluate the long-run marginal cost of expanding, maintaining and operating the distribution system. This is achieved by calculating the network cost of adding a 500 MW load to the system maximum demand, including security and the aggregated maximum demand obtained through various diversity factors applied to individual loads. The model can take into consideration a mixture of consumers specific for a particular system [7].

These costs are then allocated across voltage levels and customer groups such that the resulting DUoS charges are somewhat cost reflective. This is achieved by identifying the contribution of each customer group to the incremental distribution system cost. The resulting tariff takes the form of maximum demand and unit related charges. Maximum demand charges are used for levels of the system close to customers, on the basis that their size determines the capacity of the local network to which they are connected. These charges are expressed in terms of £/kVA month. On the other hand, unit based charges (£/kWh) reflect the impact on the network cost further up the system. This latter approach is supported by the argument that the customer individual maximum demand is less likely to coincide with the system maximum demand.

In the UK these charges vary with the voltage level, and typical ranges are given in Table 8.3. Under present practice, generators embedded into distribution networks are not subjected to DUoS charges and, as discussed in Section 8.2 above, are expected to pay their deep connection charges only. This framework is not consistent as it does not recognise the impact of operating embedded generation on network reinforcement costs. Some RECs would, in principle, allow a credit to a generator for the value of the reinforcement to the distribution network which results from the connection on the basis of an apportionment of avoided capital cost. However, there is no generally developed mechanism upon which a generator can demonstrate its positive effect on the network investment cost.

8.3.2 Contribution of embedded generation to network security

Under the terms of their licences, RECs must plan and develop their systems to the standard specified in Engineering Recommendation P.2/5 (UK security standards) [8]. The document defines the level of security of supply at each range of group demand to which the system must be planned. The level of security present is generally proportional to the resources devoted to its provision. These resources are in the form of

Table 8.3   *Typical values of DUoS charges*

DUoS charges/voltage level	Low voltage	High voltage
Availablity charge (£/kVA month)	0.99–1.8	0.94–1.5
Day unit (0700–0000) (£/MWh)	6.8–7.8	3.9–5.2
Night unit (0000–0700) (£/MWh)	1.0–1.6	0.6–1.2

investment in network facilities or generation sources. In principle, the presence of embedded generation can make a positive contribution towards local security of supply. Engineering Recommendation P2/5 defines how a mix of conventional generation plant, steam and gas turbine units, and network facilities should be treated in the context of security of supply. The basis behind the assessment of the effect of generation in P2/5 is the consideration of 'equivalence' of generation and transmission. Appendix A2 of P2/5 states that the contribution of generation can be considered to be $\frac{2}{3}$ or 67% of declared net capacity without any specific explanation. However, although P2/5 is a stand-alone document, the basis of P2/5 and the values quoted in it have a stronger and more reasoned background. Part, but not all, of this background is contained in an application report, ACE Report no. 51 [9], with reliability cost assessments provided in ACE Report no.67 [10]. An extended reasoning of effective generation is given in Appendix A3 of ACE Report 51.

ACE Report 51 defines the 'effective generation' contribution as the circuit capacity which, when substituted for the generating plant in various generation/circuit combinations, resulted in the same reliability of supply from each of these arrangements. Assuming a generating unit availability of 86% and perfect lines, an average value of 67% is then used for the ratio between effective output and maximum output of each generator. It is evident that this single value may be convenient for deterministically assessing system security, but it cannot and does not take into account the fact that the ratios themselves are variables (the report found values in the range 0.4–0.9), unit availabilities can vary significantly, and lines and transformers are not perfectly reliable.

Rules of thumb based on average availability of the conventional plant were derived for systems containing stations manned on a three, two and one shift basis. There are no definite rules defined to identity contributions of embedded generation based on generation technologies used by contemporary embedded sources.

The tacit assumption made in ACE Report 51 and Engineering Recommendation P2/5 that a certain level of load can be supported at all times is erroneous: the level of load specified is at best the average load that can be supported. However, it should be noted that in practice the actual available capacities can be higher or lower than this value. It is important to note that an average value is one parameter of a probability distribution (the location parameter) and may not necessarily be a value that can occur in reality: this is particularly the case with systems having a small number of discrete states.

In the context of their contribution to system security, generators are sometimes divided into two categories: so-called firm plant such as coal, combined and open cycle gas turbines, and non-firm plant, such as CHP schemes and renewable generation. As non-firm generation will operate in some periods at zero output this is declared to have no value to system security. It is frequently argued that, as the availability of generation is considered to be significantly lower than the availability of distribution circuits, no value can be attributed to this generation. For systems with significant penetration of embedded generation and networks with a wide diversity of primary energy sources, it is not appropriate to deny the existence of the contribution of embedded generation to the security of supply.

In the light of the above discussion, it is not surprising that there is a widespread view in the UK electricity supply industry that there is a need for a review of the security standards as they take no account of changes in the structure of the industry, of recent developments in regulation of distribution networks [11, 12], and of changes in embedded generation technology.

8.4 Allocation of losses in distribution networks with EG

The presence of EG changes the power flow patterns and therefore the losses incurred in transporting electricity through transmission and distribution networks. Embedded generation (EG) normally, but not necessarily, contributes to power loss reduction as the generation energy flow is generally against the major net flow. Loss adjustment factors (LAFs) are currently used to gross up demand/generation to the grid supply point (GSP). These factors can be used to quantify the value of the output of generators in terms of their impact on losses. The LAFs are site and time dependent. Time dependence comes from load variations and incorporates the non-linear relationship between network flows and losses. RECs normally define LAFs in accordance with the voltage level at which the load/generation is connected, and those values are used throughout the REC. Typical ranges of LAFs are given in Table 8.4.

For larger users, a substitution method is used to calculate LAFs. This method has also been recommended by the England and Wales Electricity Pool [13] to be used for allocating losses in networks with EG. In accordance with this method, the impact of a network user on the system losses is assessed by calculating the difference in losses when the user is connected and when it is disconnected from the network. There are a

Table 8.4   Typical values of loss adjustment factors

Hour/voltage level	Low voltage	High voltage
1600–1900	1.063–1.19	1.0299–1.059
0700–1600; 1900–2400	1.066–1.120	1.0279–1.050
2400–0700	1.059–1.107	1.0246–1.048

number of problems associated with this method, among which the following two give reasons for considerable concern: (i) the method can produce inconsistent results and (ii) it does not prevent temporal and spatial cross-subsidies.

These two problems of the substitution method are illustrated using the example presented in Figure 8.3. The network is composed of a single radial 11 kV feeder to which three network users are connected. A load demand of 200kW is connected at nodes A and B, while at node C there is an EG which has an output of 400 kW. Note that the length of the feeder section TA is twice the length of AB and BC.

Assuming that voltage drops and losses in the network can be neglected while calculating the network flows, approximate power flows with associated expressions for series losses are given in Figure 8.4. This approximation is adequate as the following analysis is used for comparative purposes only. Note that the approximate expression for losses, shown in Figure 8.4, is given in per unit assuming a base power of 1kW and a base voltage of 11kV.

The substitution method is now applied by disconnecting users in turn. Corresponding situations are presented in Figures 8.5–8.7.

In Figure 8.5, power flows and losses are given when the generator at C is disconnected from the network. As the total power losses increase after the EG is disconnected, in accordance with the substitution method the EG reduces losses and should be rewarded.

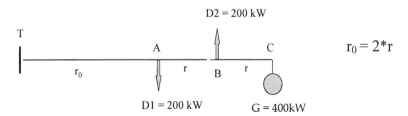

Figure 8.3   Illustration of inconsistency of the substitution method

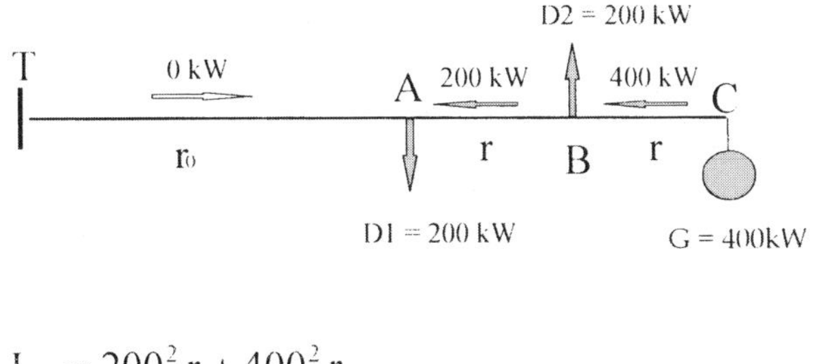

$$L_0 = 200^2 r + 400^2 r$$

Figure 8.4   Base case power flows and losses

Figure 8.6 presents the power flows and losses after the consumer at bus A is disconnected. Again, as the losses increase after the customer at A is disconnected, in accordance with the substitution method, the customer reduces losses and should be rewarded.

Finally, Figure 8.7 presents the power flows and losses after the consumer at bus B is disconnected. This is similar to the above situations since the customer at B also reduces the total system losses and could be expected to be rewarded in accordance with the substitution method.

Clearly, by applying the substitution method it appears that each of the network users contributes to a reduction in the total system losses. However, they are the only users responsible for creating losses in the sys-

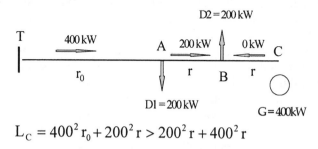

$$L_C = 400^2\, r_0 + 200^2\, r > 200^2\, r + 400^2\, r$$

Figure 8.5   *Power flows and losses when EG is disconnected*

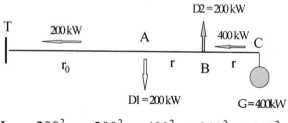

$$L_A = 200^2\, r_0 + 200^2\, r + 400^2\, r > 200^2\, r + 400^2\, r$$

Figure 8.6   *Power flows after disconnecting the consumer at A*

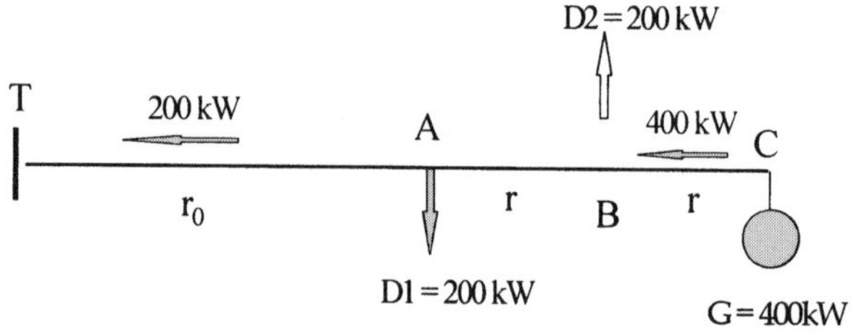

$$L_B = 200^2\, r_0 + 400^2\, r + 400^2\, r > 200^2\, r + 400^2\, r$$

Figure 8.7   *Power flows after disconnecting the consumer at B*

tem, and it is unreasonable to reward them all for their use of the network. This clearly demonstrates the inconsistency of the substitution method.

To illustrate cross-subsidies when applying the substitution method, an additional small EG is assumed to be connected at bus F as shown in Figure 8.8.

It should be noted that the addition of this small generator at bus F does not influence the losses caused by the users at A, B and C. Clearly, the new generator is responsible for losses in line TF, approximately $\Delta loss = 10^2 r$. Now for a loss reconciliation process the total system losses have to be recovered. As the user at F is the only user seen to cause losses in the system, while the others appear to reduce them, the EG at F would be left to pay, not only for its own losses but also the losses caused by the other three users as well as rewards to the users at nodes A, B and C for their apparent contribution to reduction in system losses. This, although an exaggerated example, clearly illustrates that the substitution method generates cross-subsidies.

A consistent policy for allocating series losses in distribution and transmission networks that ensures economic efficiency should be based on the evaluation of marginal contributions that each user makes to the total system losses. This method can be shown to be applicable to a fully competitive electricity market.

Marginal loss coefficients (MLCs) for allocating active power losses in distribution systems measure the change in the total active power losses L due to a marginal change in consumption or generation of active (P) and reactive (Q) power. These can be calculated as follows:

$$MLCP_i = \frac{\partial L}{\partial P_i} \quad \text{(active power related MLCs)}$$

$$MLCQ_i = \frac{\partial L}{\partial Q_i} \quad \text{(reactive power related MLCs)}$$

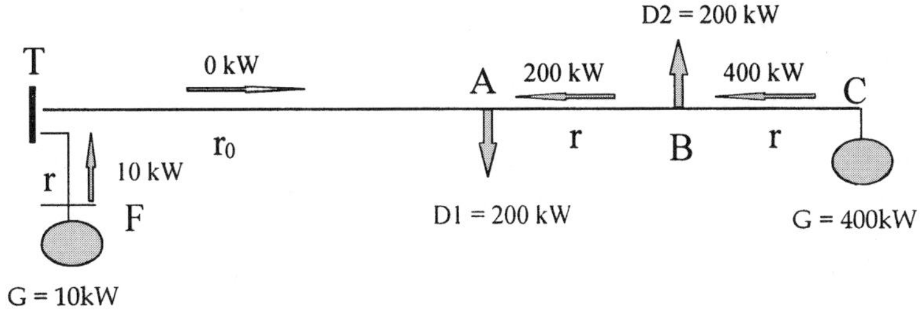

$$L_{0F} = 200^2 r + 400^2 r + 10^2 r$$

Figure 8.8   Cross-subsidies created by the substitution method

Evaluation of MLCs is relatively simple as it is an extension to standard AC load flow calculations. MLCs are fully location specific and vary in time.

Values of MLCs quantify the impact of the EG on the marginal system losses. The values of MLCs can be positive or negative, indicating whether a particular user contributes to a loss increase or decrease.

It is well known that the application of short-run marginal cost (SRMC) based pricing requires revenue reconciliation to obtain the target revenue. A similar requirement exists for MLCs. In this case the reconciliation is with reference to the total system active power losses. This is achieved through a reconciliation factor. Depending on the assumptions made in the derivation of the reconciliation factor, additive or multiplicative reconciliation can be obtained. For most situations multiplicative reconciliation is favoured [10]. The reconciliation factor (RF) can be calculated as follows:

$$\sum \left(\frac{\partial L}{\partial P_i} P_i + \frac{\partial L}{\partial Q_i} Q_i \right) RF = L$$

In practice, the reconciliation factor RF for MLCs is of the order of 0.5.

The application of the concept of MLCs on the above example can now be presented. A resistance of 0.001p.u. was used to compute the parameters used in Figures 8.9 and 8.10 that summarise graphically the impact of EG on losses and their allocation to all network users. Figure 8.9 presents the variation of total losses with EG output, whereas Figure 8.10 depicts the variation of MLCs with EG output.

The polarity of MLCs should be interpreted in accordance with the following convention:

• negative MLCs for load – load is charged for increasing the total losses
• positive MLCs for load – load is compensated for decreasing the total losses
• negative MLCs for EG – EG is compensated for decreasing the total losses
• Positive MLCs for EG – EG is charged for increasing the total losses.

If both load and generation are present at a particular node, MLCs at this node must be applied to both provided the sign convention described above is adhered to.

It is evident from Figure 8.9 that minimum losses in this network occur when the EG output equals approximately 250 kW. Beyond this level of output, EG ceases to have a beneficial effect on losses. This is reflected in the variation of the MLC at node C, to which the EG is connected (see Figure 8.10). Notice that the MLC at this node increases linearly from a negative value when EG is equal to zero and passes through zero when EG output is equal to approximately 250 kW.

Beyond 250 kW, the MLC at node C becomes positive, signalling that EG is no longer contributing to system loss reduction and is therefore required to pay for losses.

The impact of EG output on the contribution to total losses by the loads at nodes A and B is shown in Figure 8.10. The contribution of the load at node B to total losses diminishes relative to that at node A as the output of EG increases from zero to 500 kW. This trend is to be expected since node B is closer to the EG and therefore the load at this node should derive greater benefit from an increase in EG output than that at node A which is further away. As a matter of interest, the MLC profiles at node A and B cross over when EG output equals approximately 200 kW. Once EG output exceeds 400 kW, MLCs at both nodes A and B become positive, indicating that loads at these nodes should be paid for reducing losses. The reason for this is simple. EG output beyond 400 kW

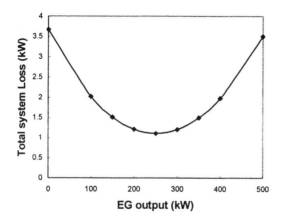

Figure 8.9   Variation of total system loss with EG output

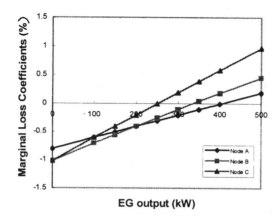

Figure 8.10   Variation of marginal loss coefficients with EG output

must be exported to the grid, causing a significant amount of losses along the route. Therefore, loads at nodes A and B, located along the way, serve to reduce the amount of power flowing to the grid and hence the marginal losses. This is why the MLCs at nodes A and B are positive, indicating that the loads at these nodes should be paid for reducing the marginal losses.

This simple example illustrates the main attributes of MLCs, in particular their ability to properly reward or penalise EG for its impact on losses.

MLCs for particular system operating points are required for hourly or half-hourly settlements. Furthermore, MLCs can also be calculated for typical days and seasons on a year-round basis to provide signals to existing and potential system users (customers, generators and suppliers) on the costs/benefits they can expect based on their impact on losses. Implementation of these marginal loss coefficients can take several forms. In the simplest case, published profiles of loss adjustment coefficients for each node and characteristic day could be used to compute the loss with the demand/generation for each network user based on either metered data or profiled consumption data. Any discrepancy between actual losses and losses computed from loss allocation coefficients could be settled through reconciliation. In this implementation scheme, it must be noted that the MLCs are computed on the basis of demand projections and therefore are subject to the same uncertainties as the load forecast.

Another, rather more accurate, approach would require the MLCs to be recalculated at each settlement period. The MLCs published in advance would therefore only serve as planning guides. In this case MLCs and associated charges would be applied on an ex-post basis. The charging regime described in the first approach can be said to be ex-ante as the MLCs are known in advance. Since the computation of MLCs is a simple task, the second option presents a workable alternative in addition to being a more equitable regime.

In the light of the UK's new trading arrangements that are currently being discussed, MLCs based on year-round analyses could be used to contract for losses in advance and only incur extra payments to cover forecast and other errors. This is an attractive option given that losses are likely to be expensive, as they will be priced at the spot price determined by the balancing market.

8.5 An alternative framework for distribution tariff development

The network operation, development and pricing frameworks are clearly of considerable consequence to the commercial performance of both network and embedded generation owners and developers. Distribution network operation and planning practices, together with adopted pricing policies, define the level of access available to participants in the electricity marketplace and therefore make a considerable impact on the amount of generation that can be accommodated. In other words, as adopted technical and commercial arrangements actually dictate the degree of openness and accessibility of distribution networks, it is vitally important to establish a coherent and consistent set of rules on both the technical and commercial fronts. It is also important to remember that distribution and transmission networks are natural monopolies and that an active involvement of the Regulator in these matters is essential.

Tariff development for distribution network services can be decomposed into two basic processes. The first process involves the determination of the allowable revenue for the network owner. Allowable revenue is a function of the costs, both operating and capital, incurred by the network owner in providing the service, including an element of profit. The second process entails allocation of the costs determined in the first process to users of the network in an economically efficient manner. Because of the monopolistic nature of the distribution business, both of these processes must be, and usually are, closely supervised by a government appointed regulatory agency whose basic mandate is to protect the public interest and ensure economic efficiency in the pricing policy. The above discussion of distribution tariffs in present use has revealed that these tariffs are not economically efficient and are not adequate to facilitate competition in generation. Radical changes in existing distribution network planning, operating as well as pricing practices are, therefore, required urgently to encourage further development of competition in generation.

In this section an alternative framework for the development of cost reflective tariffs for use in distribution systems with embedded generation is outlined. The proposed pricing regime is an extension of the concept presented in References 15 and 16, and addresses many of the deficiencies inherent in present tariff structures and provides a consistent and transparent basis for regulatory oversight of the distribution business. Following the two-step tariff development process described above, a framework for determining the allowable costs is first described. The proposed framework is based on the concept of the 'reference network'.

The concept of a reference network is a construct derived from economic theory and has a long history [15]. In economic terms the reference network is commonly referred to as the 'economically adapted network'. This latter term is somewhat more descriptive as it implies economic optimality of the network. In the context in which it is applied here, the reference network has the same topology as the existing network but, unlike the existing network, it has optimal circuit capacities. It is important to distinguish network design for pricing, as described here, from technical network design. Fundamentally, technical network design involves making decisions on network topology and on other technical issues such as voltage levels, substation layout and protection, etc., whereas network design for pricing always assumes a fixed network topology and voltage levels. It is therefore usual for technical network design to precede network design for pricing.

Determination of optimal network capacities is itself a two-stage process. The first stage determines the capacity for pure transport of electricity. The second stage determines the extra capacity required to satisfy security constraints. In both processes, the required capacities are determined using optimisation procedures with appropriately formulated objective functions. Overall, the capacities of the circuits on the reference network are determined through minimising the total network operating and capital cost as well as the cost of not delivering the service.

8.5.1 Stage 1: Optimal network capacity for transport

Optimal network capacity for pure transport of electricity is determined through an optimisation process where annuitised network capital costs and annual network operating costs (of which network variable losses are the most significant component) are traded off. This optimisation requires a calculation of the annual network cost of losses and involves modelling of annual variations of load and generation as well as associated electricity prices including the mutual correlation between these quantities. The network costs of losses are then balanced against annuitised network capital costs to determine the optimal capacity required for economic transport of electricity.

An exercise of balancing annuitised circuit capital costs and annual cost of losses was performed in Reference 17. Preliminary results, given in the form of the optimal circuit utilisation (ratio of maximum flow through the circuit and optimal circuit capacity), for different voltage levels, are presented in Table 8.5. These results indicate that the optimal utilisation of distribution circuits, particularly at lower voltage levels, should be quite low. Furthermore, the optimal design of the circuits with cost of losses being taken into account is likely to meet the security requirements (driven by the minimum design recommendations P2/5) at no additional costs in a large proportion of the system.

It is important to emphasise that the coincidence of high network loading with high electricity prices tends to result in a significant cost of losses. On the other hand, due to the maturity of the technology and the competition in manufacturing of cables and overhead lines, the cost of these products has fallen considerably over the last several years. These two effects drive distribution circuit capacities to be considerably larger than previously. Similar results are reported in Reference 18.

These findings show that losses, not peak demand, may be the main design and investment driver at low and medium voltage levels of distribution networks, particularly overhead. This may have two important implications regarding the pricing of network services with embedded generation: (i) the charges for the use of lines should be based on the user's contribution to losses, not on the individual peak load, and (ii) the availability of embedded generation may not be very important for determining its ability to substitute distribution circuit capacity. In other words, the fact that the availability of embedded generation is considerably lower than the availability of distribution circuits may not be very relevant to the potential of EG to postpone reinforcement of distribution circuits.

8.5.2 Stage 2: Security driven network expenditure

Since privatisation, Regional Electricity Companies are required both to comply with P2/5 and also to report to the Regulator on their performance against Guaranteed and Overall Standards. These standards are set to ensure a minimum level of service and to encourage higher standards. The most important network performance indices reported annually by OFFER [11] are number of interruptions per 100 customers (so called 'security') and Customer Minutes Lost (so called 'availability'). One of the main consequences of the introduction of the Standards is the change in the approach to network investment strategies which are now being more and more customer, rather than utility, driven. Similarly, the indices used to measure the company's performance are the main drivers in the development of network reinforcement and replacement strategies rather than P2/5.

This approach effectively replaces the existing deterministic security standards by probabilistic ones. To determine the additional capital and operating expenditure driven by service quality requirements, which is at

Table 8.5   Optimal utilisation factors of cables and overhead lines in a typical distribution network

Voltage level	Type of conductor	
	Cable	Overhead line
11 kV	20–40%	13–20%
33 kV	30–50%	17–25%
132 kV	75–100%	30–50%

the heart of the probabilistic standards, these costs should ideally be balanced against outage costs. The quantification of customer outage costs and customer worth of supply is discussed in Reference 19. In probabilistic security assessments, each network element, including embedded generating units, would be assigned an availability profile which contains information regarding both frequency and duration of outages (for network and generation) and frequency and duration of various levels of generation output. This would allow a comprehensive reliability study to be performed from which it would be possible to quantify the relative contribution of generation to reliability of supply. This is a very complex optimisation problem for which effective solution algorithms are yet to be developed. A framework for solving such optimisation problems is described in References 20–22.

The specific issue of the contribution of EG to network security is discussed in Reference 23. Starting from P2/5, this paper presents two alternative approaches indicating a new way forward in creating appropriate security standards in networks containing all forms of embedded generation. The approach used in ACE Report 51 and P2/5 can be improved by evaluating the expected load lost or not supported. This value can be converted into a load level which represents the expected load that can be supported and which is equivalent to the value used in ACE Report 51 and P2/5. These studies show that this value of load is less than the level specified in ACE Report 51 and P2/5. The difference is sometimes small even with unit availabilities of 0.86 (the P2/5 assumed value), becoming smaller for more modern units (say with an availability of about 0.95), but considerably larger and very significant for units with low availabilities such as wind turbines.

However, it is important to bear in mind that the concept of expected load lost is not a complete answer since this is still an average value with all the weaknesses associated with such values. An alternative concept of 'percentage of time that the support can be relied on' is therefore proposed as a much more reasonable approach since this provides an objective measure of how often the required level of security can be provided by the combination of embedded generation and lines or transformers. The capacities that can be relied on for specified periods of time can easily be deduced. The results show that the capacities that can be relied on are often much lower than the value tacitly assumed in P2/5, particularly if the required security should be available for most of the time, say, greater than 95–99% of the time. In such cases, the capacity levels can be as low as 60% of the value assumed in P2/5. This is a serious reduction and one which suggests that the concept of P2/5 should be questioned.

However, it would not be appropriate to conclude that, in the context of network performance, the value of embedded generation is negligible. Although, embedded generation may not be able to substitute for distribution circuits, due to considerable difference in availability, the presence of generation could significantly improve network performance. Consider, for instance, a load point with an average annual duration of outages of, say, 10 h. Suppose that a generator, capable of supplying this demand, is being connected at the same node as the load. Even if the availability of the generator is very low, say 50%, there would be a 50% chance that this generator would be available to supply the load in cases where the network is not available. This would mean that (theoretically) the duration of load outages (customer minutes lost) would be reduced by 50%, which could be indeed very significant.

8.5.3 Stage 3: Pricing – allocation of costs

Once the reference network is designed using the above two-stage process, the associated costs of losses and network investments can be optimally allocated to the users by employing conventional marginal costing principles. The corresponding allocation factors represent the optimal set of charges (for both losses and wires) which can be proven to reflect accurately the temporal and spatial cost streams. For some users, presumably small EG located close to large loads, charges for the use of distribution network and losses can be negative, if the way they use the network tends to reduce capacity requirements and losses on the reference network. In contrast, in areas dominated by generation, consumers of electricity may expect to receive some rewards for the use of the network. The impact of the level of EG availability on the demand for network capacity, and therefore related charges, can also be determined in this process.

The set of optimal prices calculated on the reference network is then applied to the existing network. The optimal prices not only reflect accurately the impact that each user has on the network cost, but also give an incentive to network owners to develop their networks in an optimal manner. This pricing encourages network investment up to the optimal capacity and discourages overinvestment. Clearly, the total distribution revenue could be combined with the capacity of the optimal distribution network, with the network losses and with network performance under both normal and emergency conditions. This in turn provides an ideal transparent framework in which regulatory agencies can exercise their statutory responsibilities in terms of monitoring the revenue, expenditure and performance of the distribution businesses.

Although possible, there is no fundamental reason to differentiate between connection and use of system charges. What is important is to identify the amount of network investment driven by each particular user. When this is achieved, then the connection charge could be associated with assets which are 100% used by the individual user, while the costs of other, shared assets, are then recovered through use-of-system charges.

Use-of-system charges related to the investment in network lines and transformers should also be re-evaluated after the new EG is connected, irrespective of whether reinforcement was required or not. As indicated in the above discussion, this first requires a redesign of the reference network. It is important to emphasise that the process of designing the reference network requires the identification of investment drivers. This is essential for deriving an adequate charging mechanism. For example, if the main investment driver is the cost of losses, the network investment cost should be allocated with respect to the contribution of individual users to the cost of losses. In this case use of system charges should be unit based. If, however, security requirements are the major investment driver, the network cost should be allocated with respect to the maximum demand at times of maximum flow through a particular network element. Preliminary analysis shows that the bulk of the reference network investment may be driven by losses, not by security. This could be important for the assessment of benefits and cost of embedded generation as it could make the difficult issue of the availability of EG less relevant in the assessment procedure.

The resultant network use-of-system charges are fully cost reflective and avoid both temporal and spatial cross-subsidies. Tariff complexity is no longer considered an insurmountable problem given the proliferation and capabilities of modern information technology (IT) systems. Modern IT systems are capable of processing an enormous amount of information efficiently. Problems related to tariff complexity could therefore be solved by appropriate use of IT in combination with innovative metering technologies. The concept of price zones could be applied to reduce the total number of nodes with different tariffs. Demarcation of the network into price zones would, of course, reduce the amount of data handling and somewhat simplify the tariffs as a whole. However, price zones come with a penalty in the form of blunted cost messages to network users. This, in turn, leads to a reduction in economic efficiency. These compromises and trade-offs are inevitable in the quest to satisfy conflicting objectives of economic efficiency and simplicity of tariff structures.

Also needed is a pricing strategy which would recognise the real contribution that embedded generation makes to security of supply. Clearly, from an EG perspective, the fundamental issue is the economic efficiency of network tariffs and their ability to reflect cost and benefit streams imposed by various network users. As emphasised above, the economic impact of EG on networks is very site specific, it varies in time and depends on the availability of the primary sources, size and operational practices of the plant, proximity of the load, layout and electrical characteristics of the local network. Therefore, it is not surprising that the relatively simplistic tariff structures, with network charges being averaged across customer groups and various parts of the network with no real ability to capture the temporal and spatial variations of cost streams, do not appropriately reflect the economic impact of EG on distribution network costs.

It is therefore essential to recognise that only location specific tariff regimes are candidates for adequately recognising the cost/benefit of EG to power networks. Furthermore, to reflect the fact that network operating and capital costs are known to be influenced by variation of demand and generation in time, a further requirement of the ideal pricing policy is that it must possess a time-of-use dimension. Consequently, in the development of the ideal pricing mechanism for networks with EG, the key requirements of spatial and time-of-use discrimination must rank very high. Furthermore, the tariff should be able to deal adequately with issues of generation availability, as this step may be vital when identifying the 'value' of availability of EG in terms of its ability to improve network performance.

8.6 Conclusions

Although a number of technical challenges related to the integration of embedded generation (EG) and power systems have yet to be addressed, this chapter discusses the importance of the integration of EG within a consistent commercial framework. It has been recognised that the present arrangements and mechanisms for pricing of distribution services do not treat EG adequately or systematically. The commercial issues related to charges for losses, connection and use of the distribution and transmission networks have been the subject of negotiations between local distribution utilities and EG companies, where many different arguments have been used in different contexts to address essentially very similar questions.

The technical and network pricing frameworks which are chosen are of considerable consequence to the commercial performance of both network and generation owners and developers. Distribution network operation and planning practices, together with adopted pricing policies, define the level of access available to participants in the electricity marketplace and therefore make a considerable impact on the amount of generation that can be accommodated. In other words, as adopted technical and commercial arrangements actually dictate the degree of openness and accessibility of distribution networks, it is vitally important to establish a coherent and consistent set of rules on both technical and commercial fronts.

8.7 References

1 SCOTT, N.: 'Are PES connection costs too high?'. Proceedings of the 1997 British Wind Energy Conference, 16–18 July 1997

2 European Standard EN 50160. Voltage characteristics of electricity supplied by public distribution systems, European Commission, Brussels, 1994

3 Engineering Recommendations G.59/1. Recommendation for the connection of embedded generating plant to the regional electricity companies' distribution systems, Electricity Association, Engineering and Safety Division, London, 1991

4 Engineering Recommendations G.75. Recommendation for the connection of embedded generating plant to the regional electricity companies' distribution systems above 20kV or with outputs over 5 MW, Electricity Association, Engineering and Safety Division, London, 1996

5 VDEW. Parallelbetrieb von Eigenerzeugungsanlagen mit dem Mittelspannungsnetz des Elektrizitatsversorgungsunternehmens (EVU), Ausgabe, VWEW – Verlag, Frankfurt, 1994

6 SCHWEER, A., *et al.*: 'Impact of increasing contribution of dispersed generation on the power system', CIGRE Study Committee no. 37, WG 37–23, Final Report, 1998

7 SMITH, R.: 'Distribution business charges for embedded generation'. IEE Colloquium on *Economics of embedded generation* (98/512), London, 1998

8 Engineering Recommendations P.2/5. Security standards, Electricity Association, Engineering and Safety Division, London, 1978

9 System Design and Development Committee of the Chief Engineers' Conference: 'Report on the Application of Engineering Recommendation P2/5, Security of Supply'. ACE Report no. 51, Electricity Council, May 1979

10 System Design and Development Committee of the Chief Engineers' Conference: 'Report on Reliability Investment in Radial H.V. Distribution Systems with Overhead Lines'. ACE Report no. 67, Electricity Council, March 1979

11 OFFER: Report on Distribution and Transmission System Performance 1997–1998, November 1998

12 OFFER: Review of Public Electricity Suppliers 1998–2000 Distribution Price Control Review (Consultation Paper), May 1999

13 The Electricity Pool of England and Wales: 'Guidance note for calculation of loss factors for embedded generators in Settlement'. Paper SSC1390OP, April 1992

14 SCHWEPPE, F.C., CARAMANIS, M.C., TABORS, R.T., and BOHN, R.E.: 'Spot pricing of electricity' (Kluwer Academic Publishers, 1988)

15 NELSON, J.R. (Ed.): 'Marginal cost pricing in practice' (Prentice-Hall, 1967)

16 FARMER, E.D., CORY, B.J., and PERERA, B.L.P.P.: 'Optimal pricing of transmission and distribution services in electricity supply', *IEE Proc. – Gener. Transm. Distrib.*, 1995, **142**, (1), pp. 1–8

17 MUTALE, J., STRBAC, G., JENKINS, N., and CURCIC, S.: 'A framework for development of tariffs for distribution systems with embedded generation'. CIRED '99 Conference, 1–4 June 1999, Nice

18 GREEN , P. SMITH, S., and STRBAC, G.: 'Assessment of alternative

distribution network design strategy', *IEE Proc. – Gener. Transm. Distrib.* 1999, **146**, (1), pp. 53–60

19 KARIUKI, K.K., and ALLAN, R.N.: 'Evaluation of reliability worth and value of lost load', *IEE Proc. – Gener. Transm. Distrib.*, 1996, **143**, (2), pp. 171–180

20 MUTALE, J., and STRBAC, G.: 'Security consideration in transmission pricing'. University Power Engineering Conference, Manchester, 1997

21 STRBAC, G.: 'OPF in management of system security'. IEE Colloquium on *Optimal power flow*, London, 1997

22 MUTALE, J. JAYANTILAL, A., and STRBAC, G.: 'A framework for allocation of loss and security driven investment in distribution systems'. IEEE PowerTech Conference, Budapest, September 1999

23 ALLAN, R.N. STRBAC, G., and KAY, M.: 'Security standards and contribution made by embedded generation'. To be presented at DRPT 2000 Conference, April 2000, London

Lindsey Oil Refinery Co-generation Plant
The plant produces 118 MW heat and 38 MW of electrical energy.

(Source: National Power PLC)

Chapter 9

Concluding remarks

Embedded generation offers considerable environmental benefits and so its continued development can be considered to be beneficial for society as a whole. CHP has the obvious advantage of increasing the overall efficiency with which fossil fuels are used, while the use of renewable energy sources is clearly desirable as a way of reducing gaseous emissions from conventional generating plant. It is the policy of most European governments to encourage the development of CHP and the exploitation of the new renewable energy sources. So far these technologies, generally, have been implemented in reasonably large units (>100 kW) and in penetrations that, although starting to become significant, have not yet led to major changes in existing power systems. A number of emerging technologies (e.g. fuel cells, advanced Stirling engines and micro-turbines) have the potential for cost-effective domestic CHP, while it is suggested by some that photovoltaics integrated into the fabric of houses will become economically attractive in the future. If these developments do take place, they would result in extremely large numbers of very small (< 5kW) embedded generators connected very widely through the distribution network. Thus, it is realistic to contemplate the probability of a continued increase in the capacity of embedded generation of the types currently in service (e.g. industrial/commercial CHP, wind farms, biomass plants) and the possibility of a significant increase in very small domestic scale units.

Much of this book has been devoted to a discussion of how embedded generation plant interacts with the distribution network and the technical details of the connection. It is technical matters to do with the connection which are of immediate concern to practising engineers, as they deal with embedded generation projects on a daily basis, and for a single generator these issues are fairly well understood. The effect of many embedded generators on a network is only now being addressed, and more work is required in the development of standards and practices to allow multiple embedded generators to be connected to the network safely and effectively at minimum overall cost.

The economic and commercial aspects of the connection of embedded generators remain confused. Deep charging arrangements whereby the embedded generator pays for all work required on the distribution network is clearly contentious as the charges incurred vary with the residual capacity in the network and, if more than one generator wishes to connect to a network, the order in which projects are developed.

With the increasing amount of embedded generation coming on to distribution systems, it is perhaps time to look beyond the issues of the connection of a single embedded generation plant and begin to consider the overall effect of large penetrations of embedded generation on the future of electric power systems. A significant penetration of embedded generation changes the nature of a distribution network and increases considerably the complexity of both the design and operation of the circuits. Embedded generation also affects transmission networks and central generation. In its present way of operating, in response to available renewable energy sources or heat demand, embedded generation will generally increase uncertainty in the power system and reduce the mean but increase the variance of the power required from central generating plant and the flows in transmission and distribution circuits. There are clearly costs associated with managing this additional uncertainty and complexity of operation, and mechanisms need to be found to allocate these equitably. A number of energy storage technologies are now emerging into commercial reality which may find application in embedded generation plants as a means of managing the uncertainty caused by CHP and renewable generation.

As embedded generation changes the flows in the power system it alters the operating losses. If an embedded generator is adjacent to a large load at a time of high demand on the network then it will significantly reduce network losses. Conversely, if an embedded generator is remote from large loads then at times of minimum system demand the generator will increase system losses. Calculation of the impact of embedded generators on losses is certainly possible but, as has been discussed in Chapter 8, it is not clear that the present arrangements, at least in the UK, send correct economic signals.

Electric power systems are very capital intensive and the effect of embedded generation on the requirement for conventional generation, transmission and distribution assets is extremely important. If embedded generation increases the requirements for capital assets on the network then it can be argued that these costs should be borne by the generator while, conversely, if embedded generation reduces the requirements for assets then any savings should be returned to the generator. This is a significant but extremely contentious issue as it is first necessary to determine the change in requirement for assets and then arrive at some arrangement whereby the transmission and distribution utilities, who operate a natural monopoly with the associated obligations, are able to respond to the changes in the requirement for their services of transporting electrical power. An important part of the calculation of the change in requirement for system assets is the contribution which embedded generation makes to system security. Given the stochastic nature of the output of much embedded generation it is difficult to see how this may be undertaken without using probabilistic methods.

At present most embedded generation is considered as 'negative load' over which the distribution utility has no control. Embedded generation is not integrated into the design or operation of the power system, and a worst-case scenario always has to be considered whereby any effect of the embedded generation plant will be to the detriment of the power system. The most obvious example of this segregation of embedded generation from the power system is the widespread use of sensitive loss-of-mains protection with the result that in the event of a system disturbance the embedded generation is disconnected from the network, just when generation is most needed. As long as the view persists that embedded generation is somehow apart from the rest of the power system, then the overall system, which consists of central generation – transportation facilities – embedded generation – customer load, cannot either be developed or operated in an optimal manner.

One view of possible future developments is shown schematically in Figure 9.1, which may be compared with Figures 1.4 and 1.5 at the beginning of this book. Figure 1.4 showed the traditional passive distribution network. Figure 1.5 indicates the state of the art today, with independent generators connecting to the distribution network but taking little part in its operation. Figure 9.1 illustrates how the control of embedded generators and the distribution network might be integrated

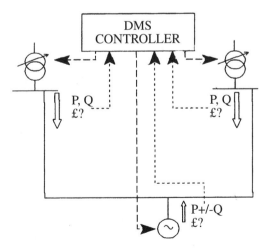

DMS Controller inputs:
Network P, Q, V,
Contracts for ancillary services
DMS Controller outputs:
Tap positions,
Generator despatch

Figure 9.1   Integrated control of distribution system with embedded generation. Only one generator shown, for clarity

in the future. This would require both a suitable commercial framework (perhaps based on contracts for ancillary services) and a reasonably sophisticated distribution management system (DMS). In transmission systems these types of commercial arrangements and control systems are reasonably well understood but they have not yet been applied to the much more extensive distribution system.

Thus, the view which is taken of embedded generation seems to be extremely important for the long-term development of the power system. It can be argued that it is only by integrating embedded generation into the power system that the overall costs of supplying customers with electrical energy will be minimised. There are, of course, serious technical considerations associated with the energy sources of the embedded generators and their location in the network However, a framework of understanding exists for many of these issues. What seems to require considerable further work is the development of understanding and mechanisms by which a large number of independent operators of embedded generation plant may be influenced to install and operate their equipment to minimise overall costs, particularly in a deregulated electric power system.

Index